EXPLORING GLOBALIZATION OPPORTUNITIES AND CHALLENGES IN SOCIAL STUDIES

A.C. (Tina) Besley, Michael A. Peters,
Cameron McCarthy, Fazal Rizvi
General Editors

Vol. 26

The Global Studies in Education series is part of the Peter Lang Education list.
Every volume is peer reviewed and meets
the highest quality standards for content and production.

PETER LANG
New York • Washington, D.C./Baltimore • Bern
Frankfurt • Berlin • Brussels • Vienna • Oxford

EXPLORING GLOBALIZATION OPPORTUNITIES AND CHALLENGES IN SOCIAL STUDIES

Effective Instructional Approaches

EDITED BY LYDIAH NGANGA,
JOHN KAMBUTU, AND
WILLIAM B. RUSSELL III

PETER LANG
New York • Washington, D.C./Baltimore • Bern
Frankfurt • Berlin • Brussels • Vienna • Oxford

Library of Congress Cataloging-in-Publication Data

Exploring globalization opportunities and challenges in social studies:
effective instructional approaches / edited by Lydiah Nganga,
John Kambutu, William B. Russell III.
p. cm. — (Global studies in education; v. 26)
Includes bibliographical references.
1. Social sciences—Study and teaching—United States.
2. Globalization—Study and teaching. 3. International education—United States.
I. Nganga, Lydiah. II. Kambutu, John. III. Russell, William B.
LB1584.E98   300.71—dc23   2012029477
ISBN 978-1-4331-2129-6 (hardcover)
ISBN 978-1-4331-2128-9 (paperback)
ISBN 978-1-4539-0964-5 (e-book)
ISSN 2153-330X

Bibliographic information published by **Die Deutsche Nationalbibliothek**.
Die Deutsche Nationalbibliothek lists this publication in the "Deutsche
Nationalbibliografie"; detailed bibliographic data is available
on the Internet at http://dnb.d-nb.de/.

The paper in this book meets the guidelines for permanence and durability
of the Committee on Production Guidelines for Book Longevity
of the Council of Library Resources.

© 2013 Peter Lang Publishing, Inc., New York
29 Broadway, 18th floor, New York, NY 10006
www.peterlang.com

All rights reserved.
Reprint or reproduction, even partially, in all forms such as microfilm,
xerography, microfiche, microcard, and offset strictly prohibited.

Printed in the United States of America

CONTENTS

Acknowledgments ix
    Lydiah Nganga

Preface xi
    William B. Russell III

Part 1. Global Issues, Trends, Policies, Practices, and Implications

Chapter One: Globalization: History, Consequences and What to Do with It 3
    John Kambutu

Chapter Two: Teachers on the Front Line of Global Migration 13
    Catherine Cooke-Canitz

Chapter Three: (En)countering the Paradox: Challenging the Neoliberal Immigrant Identity 25
    Paul G. Fitchett

Chapter Four: Immigration and *Global* Economies in the Context of Globalization 37
    Lydiah Nganga & Keonghee Tao Han

Chapter Five: Preparing Teachers for *Global* Citizenship: Perspectives from One Caribbean Island 51
    Karen Thomas-Brown

Chapter Six: Grounding Globalization: Theory, Communication, and Service-Learning 67
    Ozum Ucok-Sayrak & Erik Garrett

Chapter Seven: Global Classrooms: A Contextualized Global Education 79
    Cameron White

Chapter Eight: Institutional Internationalization: The Undergraduate Experience 89
    Linda B. Bennett

Chapter Nine: Global Citizenship and the Complexities of Genocide Education 97
    Antonio J. Castro & Rebecca C. Aguayo

Part 2. Global Issues and Innovative Instructional Practices for Teaching Global Education

Chapter Ten: Broadening Horizons: Utilizing Film to Promote Global Citizens of Character 109
Stewart Waters & William B. Russell III

Chapter Eleven: Meeting the Challenges of Implementing Global Education in a Time of Standardization 131
Mirynne Igualada & Dilys Schoorman

Chapter Twelve: Definition Devolution: Allowing Students to Redefine and Rename Citizenship and Civic Engagement 143
Emma K. Humphries & Elizabeth Yeager Washington

Chapter Thirteen: Hearing a Chorus of Voices: Globalizing the U.S. History Curriculum with Historical Empathy 155
Joseph O'Brien & Jason L. Endacott

Chapter Fourteen: Global Education for Critical Geography 169
Jason R. Harshman

Chapter Fifteen: Blogging for Global Literacy and Cross-cultural Awareness 179
Kenneth T. Carano & Daniel W. Stuckart

Chapter Sixteen: Using Storytelling and Drama to Teach Understanding and Respect for Global Values and Beliefs 197
Thomas N. Turner, Dorothy Blanks, Sarah Philpott, & Lance McConkey

Chapter Seventeen: World Tour by Bus: Teaching and Learning about Globalization by Exploring Local Places in Search of Global Connections 207
Aaron T. Bodle

Chapter Eighteen: Teaching Social Studies from a Human Rights Perspective: Professional Development in a Context of Globalization 217
Rachayita Shah, Rosanna Gatens, Dilys Schoorman, & Julie Wachtel

Chapter Nineteen: Preparing Teachers for Global Consciousness in the Age of Globalization 227
Lydiah Nganga

Chapter Twenty: Creative Pedagogies in Integrating Global Awareness in Secondary Social Studies Curricula in Teacher Education Programs and Schools 239
Toni Fuss Kirkwood-Tucker

Chapter Twenty-one: Social Studies Education in a Globalized Era: Afterword 255
Lydiah Nganga & John Kambutu

Contributors 259

ACKNOWLEDGMENTS

I express appreciation to all the educators who answered the call for manuscripts. Due to their generous contribution of quality chapters, this book is now a reality. Because these educators have shared their experiences and wisdom generously, *Exploring Globalization Opportunities and Challenges in Social Studies: Effective Instructional Approaches* is now available as a resource for global and social studies educators. Although this work is perhaps most beneficial to such educators, others will most certainly benefit because these chapter contributors have employed a variety of philosophical, theoretical and experiential lenses relative to globalization and global education.

Many thanks to the following chapter reviewers for providing meaningful feedback: Rebecca C. Aguayo, Linda Bennett, Aaron Bodle, Kenneth T. Carano, Antonio J. Castro, Catherine Cooke-Canitz, Paul G. Fitchett, Erik Garrett, Rosanna M. Gatens, Jason R. Harshman, Mirynne Igualada, Joshua Kenna, Toni Fuss Kirkwood-Tucker, Joe O'Brien, Cynthia Poole, Dilys Schoorman, Rachayita Shah, Daniel W. Stuckart, Karen Thomas-Brown, Thomas N. Turner, Ozum Ucok-Sayrak, Stewart Waters and Cameron White. I am deeply indebted to these reviewers and chapter contributors because their efforts were on a voluntary basis.

I am equally indebted to my co-editors, John Kambutu and William B. Russell III. This book project would not have been possible without their support and expertise.

Finally, many thanks to Michael A. Peters, Chris Myers, Stephen Mazur, Phyllis Korper, Sarah Stack, and Jackie Pavlovic of Peter Lang Publishing for facilitating the publication of this book.

Lydiah Nganga
University of Wyoming, Casper College Center

PREFACE

What is global education? What is the role of education in a global world? How do we prepare students to become effective citizens in a global community? How do we effectively teach global education? These questions have spawned many debates and scholarly discussions and are not easily answered. No matter if you align with the core constructs of global interconnectedness or perspective consciousness, global education curriculum and pedagogy form an important educational concept that has and will continue to arouse interest and ignite discussion.

Exploring Globalization Opportunities and Challenges in Social Studies: Effective Instructional Approaches examines global education and how it should be taught in the social studies. This volume includes twenty-one chapters and is divided into two sections. Part 1, Global Issues, Trends, Policies, Practices, and Implications, includes chapters that examine relevant global issues. This section begins with a chapter by John Kambutu that examines globalization, its history, impact, and role in education. The following chapter by Catherine Cooke-Canitz analyzes the challenges of teacher acculturation and presents four main orientations to help in understanding the various responses that teachers have toward their immigrant students. Chapter 3, by Paul G. Fitchett, explores the dynamics of neoliberalism, its impact on immigrant identity, and the manner in which social studies textbooks and curricula have propagated neoliberal principles. Fitchett's chapter is followed by Lydiah Nganga and Keonghee Tao Han's examination of immigration issues related to economic globalization and how these issues call for K–12 teachers and university instructors to become well informed and proactive global multicultural educators. Chapter 5, by Karen Thomas-Brown, analyzes perspectives from a Caribbean island on preparing teachers for global citizenship. The following chapter by Ozum Ucok-Sayrak and Erik Garrett offers service-learning as a pedagogical practice for preparing students to be global citizens. Chapter 7, by Cameron White, discusses global classrooms and how to contextualize global education in a contemporary classroom. White's chapter is followed by Linda B. Bennett's examination of various higher education institutions' undergraduate academic international/global experiences. This section concludes with a chapter by Antonio J. Castro and Rebecca C. Aguayo, which discusses the complexities of genocide education and global citizenship.

Part 2, Global Issues and Innovative Instructional Practices for Teaching Global Education, includes chapters that explore innovative instructional practices to teach effectively to specific global issues. This section begins with a chapter by Stewart Waters and William B. Russell III that examines how films can be used to facilitate relevant and powerful discussions in the classroom to encourage the development of global citizens of character. Waters and Russell's chapter is followed by Mirynne Igualada and Dilys Schoorman's discussion of the challenges facing the teaching of global education in a time of standardization. The following chapter, by Emma K. Humphries and Elizabeth Yeager Washington, examines a conceptual framework through which educators can help students to understand the terms "citizenship" and "civic engagement." Chapter 13, by Joseph O'Brien and Jason L. Endacott, explores how the U.S. history curriculum can be globalized with historical empathy. Chapter 14, by Jason R. Harshman, emphasizes global education and the importance of including multiple perspectives when teaching cultural, human, and physical geography. In chapter 15, Kenneth T. Carano and Daniel W. Stuckart explore using Weblogs, or blogs, as a pedagogical tool in the development of global literacy and cross-cultural awareness. This chapter is followed by Thomas N. Turner, Dorothy Blanks, Sarah Philpot, and Lance McConkey's exploration of using storytelling and drama to teach understanding and respect for global values and beliefs. Chapter 17, by Aaron T. Bodle, discusses teaching and learning about globalization by exploring local places in search of global connections. Bodle's chapter is followed by Rachayita Shah, Rose Gatens, Dilys Schoorman, and Julie Wachtel's analysis of professional development training of human rights violations and its context to globalization. Chapter 19, by Lydiah Nganga, explores a variety of instructional strategies for social studies method courses to help students gain a better understanding of the world. Nganga's chapter is followed by Toni Fuss Kirkwood-Tucker's discussion of creative pedagogies for teacher candidates in teacher education programs and practicing teachers in schools, helping them to effectively teach about the world and its people despite extant state and district curricular guidelines. This section concludes with a chapter by Lydiah Nganga and John Kambutu, summarizing the role of social studies and global education in a globalized era.

The twenty-one chapters included in this volume represent the vast of amount of relevant topics related to global education in social studies. This volume serves as both a foundation and a springboard for dialog, scholarship, curriculum, and pedagogy as global education and social studies education move forward into the twenty-first century.

William B. Russell III
University of Central Florida

PART 1

Global Issues, Trends, Policies, Practices, and Implications

CHAPTER ONE

Globalization: History, Consequences and What to Do with It

John Kambutu

Globalization is an old phenomenon. However, the impact of contemporary globalization efforts is generally misunderstood due to positionality. While people in positions of benefit tend to have favorable perceptions, the disadvantaged are likely to question the value of globalization. Thus, to fully understand globalization, this chapter recommends a holistic analysis from a critical theoretical framework. In addition, a call is made for an education for social justice.

Globalization has had an impact on everyone in some way. Nevertheless, its history and consequences are somehow murky. Because globalization has different effects, people use various lenses and metaphors to understand it. For example, groups that benefit the most might use favorable metaphors such as the "global village, the Network of interdependence, the McWorld and the Spaceship earth." But the disadvantaged might make meaning through critical schemas such as "military competition, and Neo-colonialism" (Sleeter, 2003, pp. 3-4). So, while positionality shapes understanding, it also socializes individuals into a particular kind of thinking, feeling and acting (Kambutu, Rios & Castañeda, 2009). Therefore, to understand the effects of globalization, a holistic and objective analysis is essential.

The origin of globalization is a contested issue. While some scholars see it as a new phenomenon, others think globalization is as old as humanity (Wiarda, 2007). In support of the evolutionary nature of globalization, Held, McGrew, Goldblatt and Perraton (1999) provided the following four phases of development: a) pre-modern, 900 to 1000; b) early modern, 1500 to 1850; c) modern, 1850 to 1945; and d) contemporary, 1945 to the present. Advances in technology and increases in information accessibility are hallmarks of contemporary globalization.

Modern technologies and the ability to access information with relative ease have transformed the world in ways never seen before. Most notable, however, is the rise of the Net-Generation, a group that has grown up entirely in the digital age. The Net-Generation is unique in that all they know is a technologically interconnected world (Tapscott, 2009). To them, the world is "virtual," a place of interdependence and interconnections, vis-à-vis a physical place governed by rigid cultural, economic and political boundaries. Krieger (2005) offered a similar assessment but also associated globalization with improved standards. Due to interconnection, a "global village" that supports international trade and economic stability has emerged (Armijo, 1996). In addition, technologies have increased interactions between cultural groups, leading to global cultural understanding and appreciation (Kambutu & Nganga, 2008). Notwithstanding the benefits, critical theorists see a link between globalization and increasing global social injustices.

Critical theory supports objective and thorough investigation of situations. For instance, Marcuse (1973) cautioned against focusing on facts only; one must also provide objective and detailed evaluations. Thus, examining globalization using a critical theoretical framework invites a holistic investigation of the policies and practices involved (el-Ojeili & Hayden, 2006). Obviously, globalization is linked to the "global village" mindset. Supporters of a global village metaphor assign to globalization positive effects because it has enabled people of different cultural, economic and political persuasions to interact, understand and appreciate one another (Ette, 2012). Nevertheless, critical theorists hold globalization responsible for increases in global injustices (Sleeter, 2003). Indeed, designed carefully by the ruling Western elites, contemporary globalization is a framework that advances neoliberal political, social and economic agendas (Miller, 2010; Steger, 2009).

Consequences

As an invention of neoliberal Western elites, globalization focuses more on the economic, social and political interests of wealthy nations. To that end, Lee (2012) discussed globalization in the context of Anglo-merchant political, social and economic hegemonies at the expense of human and civil rights. Several strategies, including the internationalization of U.S.-based politics, economics, academic practices and goods such as movies, music and literature, create an ideal globalization climate. In addition, the popularization of global consumerism or the "McWorld" phenomenon and the subordination of poor nations' cultural, economic and political institutions through military domination support globalization efforts by spreading Western supremacy and dominance (Barber & Schulz, 1996). Notwithstanding the general impact, globalization is felt differently based on positionality.

Based on positionality, globalization affects people differently. For example, while groups in positions of advantage receive its positive effects, the disadvantaged have negative experiences. Thus, the two groups are likely to understand globalization differently. Therefore, any objective study should consider positionality. While wealthy countries that benefit the most understand globalization through the metaphors of the "global village and the Network of interdependence," the exploited groups use the "military competition, and Neo-colonialism" lenses to critique globalization efforts (Sleeter, 2003, pp. 3-4). Notwithstanding the use of different schemas, the issue of global interconnectedness is commonly held. Increasingly, however, the exploited groups hold globalization responsible for the emerging modern forms of colonialism and global injustices, particularly in education.

Effects on Education

Because of globalization, a uniform "global" curriculum has emerged. Thus, Grant and Grant (2007) were concerned about the rise of monolithic global epistemologies that mimic Western cultures. Under globalization, Western canons, American ideologies and learning structures specifically are popularized by neoliberal policies as the "norm," inherently superior, and, therefore, worthy of pursuing globally. While global curricula might have value, Preskill (2001) was apprehensive because of the potential danger of creating a false sense of global epistemological equality, while restricting "other" ways of knowing. In other words, imposing foreign educational practices on "others" could not only stifle cultural and ethnic groups' abilities to develop relevant epistemologies but also promote mental "enslavement" and global zombification.

Education for mental enslavement is dangerous. According to Woodson (1990), such an education ensures that the enslaved mind is always thinking and acting according to the enslaver's interests. So, if globalization is a framework that serves the interest of the ruling elites, a global curriculum designed by the same group should be expected to socialize world cultures into accepting their globalization policies. Consequently, Lee (2012) cautioned against neoliberal policies that support Eurocentric canons because they promote conformity instead of informed and critical discourse. Meanwhile, the privileging of the English language over various local languages is an additional disempowering policy to non-native English speakers, particularly in scholarship.

Increasingly, globalization requires scholars to write for Thomson Reuters-based academic journals, a daunting challenge for non-native English speakers. While these scholars engage in laborious educational tasks to serve an external clientele, they fail to address valuable local epistemologies. But because globalization promotes the interests of wealthy nations, it is most probable that local educational systems will be suppressed. Thus, Cabrera, Montero-Sieburth and Trujillo (2012) spoke strongly against educational systems that pretend to protect human rights while actually promoting the interests of hegemonic groups. Rather, an education that examines globalization fully in the context of social justice is necessary. But the implementation of an education for global justice might not be an easy process because globalization affects people differently. Recall that beneficiaries, that is, wealthy nations, perceive globalization favorably. Naturally, then, these nations are likely to implement curricula that do not challenge the socio-cultural, political and economic injustices supported by globalization policies (Giroux, 2006). An educational system that fails to address social justice issues could cause mental enslavement and global zombification.

An education for enslavement and zombification limits people's thinking abilities. Instead of analyzing situations critically, enslaved minds are likely to mimic the ethos of the groups in power and privilege. So, although globalization serves the interests of wealthy nations, neoliberalism has effectively popularized the notion that it is beneficial to the whole world. To prove this view, I surveyed six randomly selected people in the United States. On a piece of paper, the participants described globalization. All six participants believed in the positive nature of globalization. For example, while one respondent defined it as "marketing of products worldwide," another viewed globalization as "expanding resources to the entire world." Other participants added that globalization was simply "working together with all nations to share opportunities with Third World countries in order to create a better world."

Apparently, globalization has socialized or educated the masses in the United States into believing that it serves the interests of poor nations. As a result, the participants in my survey defined globalization favorably because it is responsible for global interconnection and improved standards. While the participants seemed to mimic neoliberal policies, they were obviously unaware of the link between globalization and increasing social injustices. As a result, groups that benefit the most from globalization might need an education that awakens a critical consciousness concerning issues of social justice. However, groups in power and privilege are likely to resist strongly an education for social justice, preferring instead an education that supports a "global village" mindset.

A global village mindset favors globalization. For example, the ongoing increased global interconnectedness caused by advances in technologies is frequently touted as evidence of improved world standards (Kambutu & Nganga, 2008). Because of global interconnection, the thinking goes, nations are now supporting one another just as individuals do in a typical village—the "it takes a village" ideology. Indeed, participants in my survey favored the global village notion. For example, one participant described globalization as the "intermixing of nations to create homogeneous cultures that collaborate and share opportunities." Although all the participants believed globalization was beneficial, they were apparently unaware of emerging social injustices such as cultural domination, displacement of people from ancestral lands, contested emigration and immigration, racism, ethnic prejudice and religious intolerance. These injustices are linked to globalization policies (Bauman, 2004; Gibson, 2010; Suarez-Orozco & Sattin, 2007).

Effects on Poor Countries

Understanding globalization through the mindset of increased cultural understanding and appreciation seems to be a worthy course. However, such a limited view fails to examine the social injustices involved. For example, although neoliberalism refers to poor countries as the main beneficiaries, wealthy nations profit the most economically, politically and socially. To that end, the World Bank and the International Monetary Fund (IMF) require the implementation of "structural adjustment programs" as a condition for granting loans to poor Third World countries (Steger, 2009). Theoretically, structural adjustment programs guard against loan delinquencies. In practice, however, the IMF's programs have had the unintended consequence of forcing poor countries into new forms of colonialism. The most notorious of the IMF's policies delves into commerce liberalization, abolition of import licensing and the reduction of tariffs. Although intended to ensure loan payments, these measures have incentivized companies from developed nations to "invest" in Third World countries (Steger, 2009).

Pressed with the pressure of paying off increasing foreign debts (debt for Third World countries grew from $618 billion U.S. dollars in 1980 to $3.3 trillion in 2007), poor countries welcome foreign investors to help raise essential funds to service external debt as required by the IMF and World Bank. As the paying off of foreign debts takes precedence, other critical services are neglected. Therefore, people in poor countries are experiencing "fewer social programs and educational opportunities, more environmental pollution, and greater poverty for the vast majority of people" (Steger, 2009, p. 55). After seeing firsthand the effects of globalization on a poor country (Kenya), a senior in my university summer international cultural program expressed his cognitive dissonance:

> Prior to this experience, I viewed globalization positively. However, it was disheartening to see greenhouses that stretch for miles through some of the most fertile soil on the planet (land surrounding Lake Naivasha), constructed by international companies

to grow flowers for export to benefit a few already privileged people while the food that should be grown there to feed the local people is imported. Although I was already aware of the inefficacy of organizations like the International Monetary Fund (IMF) to enact genuine positive change in the countries receiving aid, being able to see evidence makes it real, undeniable and frustrating. This to me is evidence that there is need to re-examine the existing relationships between developed and less developed nations, particularly in the context of globalization. As the world becomes a village, it is essential that we all become empathic stewards of its resources. Our world has enough resources for everyone. A just sharing of those resources could ensure that the basic needs of all the people on earth are met. (Field notes, June 2010)

The seizure of productive lands by international companies is a hallmark of globalization in Third World countries. As foreign companies invade productive agricultural lands, the locals are displaced and left without economic means to support themselves and their families. As a result, they seek employment with the same foreign companies occupying their lands. In poor countries, workers in foreign companies typically earn one U.S. dollar a day. Clearly, such a wage is not enough to meet a family's basic needs. Therefore, many people migrate to foreign countries in search of better economic opportunities. Spain's Canary Islands form one such immigration destination.

Effects on Economic Immigration

Perhaps due to proximity to Africa, Europe and America, the Canary Islands have historically served as an immigration destination. However, globalization policies have increased immigration to the islands to levels never seen before (Cabrera et al., 2012). While wealthy Europeans migrate in search of exotic and natural beauty, the poor from Third World countries migrate to the Canary Islands for economic reasons. However, the number of immigrant schoolchildren grew by 279 percent between 1999 and 2000 (Cabrera et al.). Although adult immigrants experience many challenges, schoolchildren have particularly daunting experiences. Cabrera and colleagues reported that immigrant students experienced devastating incidences of discrimination and racism. In one instance, a student from Senegal described school experience as painful and disorienting because, although teachers were generally accepting, other students were mean to her. They assumed that just because she was black, she was intellectually incapable. Evidently, globalization has negatively affected educational experiences for this immigrant student as well as many others around the world. As a result, a critical examination of globalization is justified.

How to approach globalization

To be sure, globalization has had positive results. For example, as an exchange of goods, it has increased international trade (Haugen & Mach, 2010). A reduction of global commerce barriers due to pressure from the IMF has provided further opportunities for increased commerce (Stone, 2002). Meanwhile, improved technologies have facilitated interactions between cultural groups, thus promoting a degree of cultural understanding and appreciation. Indeed, according to one participant in my survey of U.S. citizens, although globalization is seen in the context of "global warming, it is also about increasing knowledge of people, society, diversity and the world." So, globalization involves "increased knowledge of the world in order to promote cultural understanding and acceptance." Other respondents conceded that globalization should be understood broadly because it affects the world's "issues, business/finance and communication cultures." Further, these participants cautioned that making meaning of globalization "depends on where we (our thoughts) are," that is, it is a factor of positionality.

Globalization has both positive and negative impacts. To fully grasp the meaning, therefore, it should be analyzed carefully, objectively and critically using a critical theory framework. In addition to examining the positives, it is also necessary to use a human rights lens to explore the link between globalization and increasing social injustices because as Lee (2012, p. 133) argued, a focus on "human rights" is essential because "regardless of social differences, people yearn to connect with one another and understand deeply shared struggles. When one of us falls, we all fall." The challenge for educators and policy makers is to ensure that globalization is studied holistically within a social justice framework

Globalization and Social Justice

Because of positionality, different people understand social justice differently. For example, Pelzer (2010) reported that people in positions of power and privilege, see, albeit erroneously, social justice as a code word for forced transfer of wealth. But Nieto and Bode (2012) had contrary views. Instead, they postulated that social justice stands for human equality in the context of equity. Therefore, social justice is an effort to understand and address social inequalities and oppression from historical, economic, political and cultural perspectives. An education with a social justice tilt is, therefore, necessary in order to examine the "broader setting or the context or the culture or the institutional policies and practices" that allow individuals and society to support social inequalities based on human differences both natural and socially constructed (Pelzer, 2010, para. 23). Additionally, an education for social justice supports instructional approaches that are culturally inclusive, respectful, appreciative and sensitive to issues of justice for all. Zajda (2010) offered a similar description that highlighted social justice as an intellectual exercise that invokes passion, courage, and the spirit of human potential for a just and open world.

An education for social justice is holistic and involving. It is not business as usual. Rather, an education for social justice seeks to question and to deconstruct all areas of societal oppression and domination. Indeed, a social justice education empowers by encouraging activism and refusal to admit that conditions are "the way they are because they cannot be different," (Freire, 1997, p. 36). Therefore, an education for social justice seeks to improve and transform all lives. But effective transformation is not possible in an education for domestication. Rather an education for critical consciousness, implemented by educators with a clear understanding of the sociopolitical nature of education is essential. Additionally, educators for critical consciousness or social justice are committed to change and are willing to engage and become activist in movements of social transformation (Zajda, 2010; Nganga & Kambutu, 2009; Banks, 2007; Oakes & Lipton, 2007, Apple, 2004). A social justice educational framework, therefore, calls for transformative actions that advance equity, equitable access to education, freedom from discrimination, and the principles of a democratic society (Apple, 2010).

An education that studies globalization in the context of social injustice, global democracy and pluralism is a must. In the following chapters, educators have shared a variety of effective instructional strategies. The shared pedagogies should create essential space to deconstruct globalization critically in order to ensure justice for all.

References

Apple, M. W. (2004). *Ideology and curriculum*. New York: Routledge.
Apple, M. W. (2010). *Global crises, social justice, and education*. New York: Routledge.
Armijo, L. E. (1996). Inflation and insouciance: The peculiar Brazilian game. *Latin America Research Review*, 31(3), 7–46.
Banks. J. (2007). *Educating citizens in a multicultural society*. New York, NY: Teachers College Press.
Barber, B. R., & Schulz, A. (Eds.). (1996). Jihad vs. McWorld: *How globalization and tribalism are reshaping the world*. New York: Ballantine.
Bauman, Z. (2004). *Wasted lives: Modernity and its outcasts*. Cambridge: Polity.

Cabrera, L., Montero-Sieburth, M., & Trujillo, E. (2012). Window dressing or transformation? Intercultural education influenced by globalization and neoliberalism in a secondary school in the Canary Islands, Spain. *Multicultural Perspectives*, 14(3), 144-151.

el-Ojeili, C., & Hayden, P. (2006). *Critical theories on globalization*. New York: Palgrave Macmillan.

Ette, U. E. (2011). *Nigerian immigrants in the United States: Race, identity and acculturation*. New York: Lexington Books.

Freire, P. (1997). *Pedagogy of the heart*. New York: Continuum.

Gibson, M. L. (2010). (Are we "Reading the World"? A review of multicultural literature on globalization. *Multicultural Perspectives*, 12(3), 129-137.

Giroux, H. (2006). Reading Hurricane Katrina: Race, class and the biopolitics of disposability. *College Literature*, 33(3), 171-196.

Grant, C., & Grant, A. (2007). Schooling and globalization: What do we tell our kids & clients? *Journal of Ethnic and Cultural Diversity in Social Work*, 16(3/4), 213-225.

Haugen, D., & Mach, R. (2010). *Globalization*. New York: Gale.

Held, D., McGrew, A., Goldblatt, D., & Perraton, J. (1999). *Global transformations: Politics, economics and culture*. Cambridge: Polity.

Kambutu, J., & Nganga, L. (2008). In these uncertain times: Educators build cultural awareness through planned international experiences. *Teaching and Teacher Education*, 24, 939-951.

Kambutu, J., Rios, F., & Castañeda, R. C. (2009). Stories deep within: Narratives of U.S. teachers of color from diasporic settings. *Diaspora, Indigenous, and Minority Education*, 3(2), 96-109.

Krieger, J. (2005). *Globalization and state power*. New York: Pearson/Longman.

Lee, P. (2012). From common struggles to common dreams: Neoliberalism and multicultural education in a globalized environment. *Multicultural Perspectives*, 14(3), 129-135.

Marcuse, H. (1973). *Studies in critical pedagogy*. Boston: Beacon Press.

Miller, R. W. (2010). *Globalizing justice: The ethics of poverty and power*. New York: Oxford University Press.

Nieto, S. & Bode, P. (2012). *Affirming diversity: The sociopolitical context of multicultural education*. New York: Pearson.

Nganga, L. & Kambutu, J. (2009). Teaching for democracy and social justice in isolated rural settings: Challenges and pedagogical opportunities. In S. Greonke & A. Hatch (Eds), *Critical pedagogy and teacher education in the neoliberal era: Small openings* (pp. 191-204). Milton Keynes, U.K: Lighting Source UK Ltd.

Pelzer, J. (2010, September, 20). Ayers controversy thrusts University of Wyoming Social Justice Research Center into spotlight. *Casper Star-Tribune*. Retrieved from http:// trib.com/news/state-and-regional/article_670d08a2-dc06-539b-a795 696a018263fb.html#ixzz20X0nTjC5

Preskill, S. (2001). Contradictions of domestic containment: Forestalling human development during the Cold War. In K. Graves, T. Glander, & C. Shea (Eds.), *Inexcusable omissions: Clarence Karier and the critical tradition in history of education scholarship* (pp. 181-194). New York: Lang.

Sleeter, C. (2003). Teaching globalization. *Multicultural Perspectives*, 5(2), 39.

Steger, M. B. (2009). *Globalization: A very short introduction*. New York: Oxford University Press.

Stone, R. (2002). *Lending credibility: The IMF and post-communist transition*. Princeton, NJ: Princeton University Press.

Suarez-Orozco, M., & Sattin, C. (2007). Wanted: Global citizens. *Educational Leadership*, 64 (7), 58-62.

Tapscott, D. (2009). *Grown up digital: How the Net Generation is changing your world*. New York: McGraw-Hill.

Wiarda, H. (2007). *Globalization: Universal trends, regional implications*. Lebanon, NH: University Press of New England / Northeastern University Press.

Woodson, C. G. (1990). *The mis-education of the Negro*. Nashville, TN: Winston-Derek.

Zajda, J. (2010). *Globalization, education, and social justice*. Melbourne, Australia: Springer.

Zajda, J., Biraimah, B. & Gaudelli, W. (2008). *Education and inequality in global culture (eds.)*. Dordrecht, The Netherlands: Springer.

CHAPTER TWO

Teachers on the Front Line of Global Migration

Catherine Cooke-Canitz

> No cultural group remains unchanged following culture contact; acculturation is a two-way interaction, resulting in actions and reactions to the contact situation.
> David L. Sam and John W. Berry, 2010

Increased student diversity in the classroom is observable in today's schools. In 2009, nearly one-quarter (23.8 percent) of the 70.9 million children in the United States under the age of seventeen had at least one immigrant parent (Batalova & Terrazas, 2010). Between 1993 and 2000, the student population in the United States rose by 12 percent, while the population of students with limited English proficiency (LEP) rose by 84 percent (Migration Policy Institute, 2012). Immigrant students bring with them the hope and excitement of increasing and strengthening global networks, of expanding thoughts and experiences, and of contributing knowledge and talents to communities and lives. However, schools feel the brunt of this societal change, and classroom teachers on the front line bear the hourly, ongoing challenges associated with the language differences, cultural adaptations, and potential cultural conflicts that immigrant students often bring. Schools and teachers must see "the new composition of their student body as a turning point calling for qualitative change and the development and implementation of new coping strategies" (Horenczyk & Tatar, 2002, p. 437). To embrace the benefits of cultural globalization, teachers are called upon to confront their own prejudice, bias, and perceived threats to their security, as well as to digest the prospect of adapting, adjusting, and coping every day over extended periods of time.

Before teachers can teach in ways that promote social justice and global citizenship, they must negotiate their own acculturation. From the fields of anthropology and psychology, the term "acculturation" refers to an interactive process of change between the "host" and "other" cultures encountering one another over the long term (Berry, 2008). Although acculturation is more generally applied to nations and their policies toward immigrants, Berry (2005) states that acculturation is "equally relevant for national policies, institutional arrangements . . . and for individuals in the larger society" (p. 711). The institutions in this case are school.

Acculturation in the educational context involves teachers' and students' psychological attitudes toward acculturation, their various cultural anthropologies and biases, and the shifting dynamics of power and authority within the identity politics of the classroom. Understanding the acculturation process is vital to building a functioning and supportive classroom culture and may alleviate what Tatar and Horenczyk (2003) call "diversity-related burnout," which occurs when teachers become overwhelmed with the additional affective, behavioral, and cognitive demands placed upon them in interactions with immigrant students.

While some researchers attend to the psychological and cultural change in immigrant students in the classroom, change occurs for members of the host community as well (Horenczyk, 1997; Sam & Berry, 2010). Few research studies focus on the complexities and stresses accompanying teachers' acculturation or the ways they adapt psychologically and culturally to the immigrant students in their classrooms. Interactions with culturally diverse students challenge and shape teachers' own identities, which can either facilitate or hinder their work with students.

This chapter addresses challenges of teacher acculturation and presents four main orientations, as identified by Berry (1997), to help understand the various responses that teachers have toward their immigrant students. These orientations can empower teachers to monitor and adjust the nature of their interactions with their immigrant students. In addition, I discuss how the Interactive Acculturation Model (Bourhis, Moise, Perreault, & Senecal, 1997) predicts the probable outcomes of interactions between teachers and students, based on their selected orientations. Finally, I discuss the implications of those outcomes on teacher acculturation and the tenor of the classroom. This chapter offers considerations for classroom teachers facilitating the creation of positive classroom environments.

Acculturation and Challenges in the Public School Classroom

Since the reality of today's schools is one of increasing "cultural mismatch between teachers and their students" (Cockrell, Placier, Cockrell, & Middleton, 1999, p. 351), a better understanding of acculturation arms teachers with the ability to analyze and make informed decisions about developing constructive interactions with students, for everyone's well-being.

Acculturation and Teacher Stress

Colleen Ward (2001) synthesized acculturative changes into three main areas: affective, behavioral, and cognitive (labeled the "ABCs of Acculturation") and associated them with empirical approaches: stress and coping, culture learning, and social identification. Ward's study (2008) gave examples of how ABC pertained to tourism: affectively, she cited health and well-being; behaviorally, she referred to the processes by which people acquire the culturally relevant skills to interact effectively across cultural lines; and cognitively, she listed "stereotypes, contact, impacts, perceived threats, and intergroup relations in the context of tourism" (p. 111). Although acculturative processes in tourism are less long lived than in the classroom, the same considerations evident between natives and tourists might pertain to teachers as hosts of immigrant students. For instance, hosts also require culturally relevant skills to interact effectively across cultural lines, and hosts deal with their own stereotypes, contact, impacts, perceived threats of tourists, and intergroup relations in their particular context (e.g., classrooms).

Hosts can also suffer ill health from the stress of acculturating. Maslach (2011), an acknowledged expert on burnout, reconfirmed earlier findings that one of the results of burnout is ill health. When emotional expenditure in the classroom becomes overwhelming, teachers can experience emotional exhaustion. Because of emotional fatigue, teachers can become somewhat indifferent to their students, in an attempt to distance themselves from the emotions of constant negotiated interactions and social conflict (Maslach, 2003), a process called depersonalization. Depersonalization, in its turn, can reduce effectiveness in the classroom, as teachers withdraw from active engagement with students, leaving teachers with a feeling of lack of accomplishment, further adding to their emotional exhaustion, and completing the cycle of the burnout syndrome (Chang, 2009). This daily stress can result in health difficulties and exit from the profession.

Typically, hosts' acculturative changes are less pronounced than immigrants' because of the hosts' power in relation to other cultures (Navas et al., 2005, p. 31) and because immigrants are faced with more acculturation issues on a daily basis than are hosts (VanOudenhoven & Hofstra, 2006, p. 794). However, particularly in contexts in which there is diminished influence of the dominant culture, such as is found in cosmopolitan areas such as New York City, where teachers work with extremely diverse cultural communities, acculturation can force teachers to engage in "re-evaluation of their group identification, re-evaluation of the value of their major life events, and redefinition of their cultural identification" (Horenczyk, 1997, p. 37). This process can be threatening and emotionally taxing to individuals, including teachers.

Any social environment is a complex web of communication. Culture helps sort out social stimuli into patterns for belonging and reacting, for navigating interactions with others. When encountering people from other cultures, teachers' (and students') usual patterns of reaction and interaction may not work the same way, causing disequilibrium and stress. However, according to the process model of the integrative communication theory, stabilizing oneself by adjustment and adaptation leads to psychological shifts and growth in intercultural transformation and reinventing oneself to fit new contexts (Kim & McKay-Semmler, 2012).

Berry's Model of Acculturation

Traditionally, teachers in public school classrooms have most often used an assimilationist view when accepting immigrant students into the classroom, expecting immigrant students to relinquish their cultural identity and replace it with the host culture's language and values (Horenczyk & Tatar, 2002). However, in today's society, subtractive types of acculturation that deny heritage lead to conflict, rebellion, rejection, and other types of acculturation outcomes as immigrants and teachers interact and encounter adaptation problems (Phinney, Horenczyk, Liebkind, & Vedder, 2001). In a longitudinal research study of Arab Australian students in Australia, Kamp and Mansouri (2010) found that when students felt their teachers had no interest in them as a different culture, they disengaged from the learning process in school. As American classrooms continue their transition to diverse microcosms of globalism, American teachers run the risk of becoming marginalized in their own classrooms if assimilation is their only view toward immigrants and acculturation.

However, assimilation is not the only acculturation strategy available when teachers encounter immigrant students in their classrooms. Immigrants evaluate whether it is worth maintaining their heritage culture and how much value there is in seeking relationships with other groups.

Based on immigrants' answers, Berry (2005, p. 706) identified the following four orientations for the host culture:

> *Integration* values the maintenance of the heritage culture and adoption of certain features of the host culture.
> *Assimilation* expects immigrants to relinquish their cultural identity and replace it with the host culture.

Segregation allows immigrants to maintain their heritage culture, but does not allow them to adopt or change the host culture.

Exclusion refuses to allow immigrants to adopt features of the host culture and is also intolerant of the heritage culture.

At the same time that teachers select and implement their acculturation strategy, immigrant students select theirs, which research shows is tied, in part, to the strategy chosen by the teacher (VanOudenhoven, Ward, & Masgoret, 2006). In particular, discrimination is linked with certain orientations, which in turn affect acculturation and socio-cultural adaptation (Sam & Berry, 2010). Berry (1997, p. 9) identifies the following orientations for immigrants:

Integration desires to maintain key features of the heritage identity while adopting aspects of the host culture.

Assimilation relinquishes own cultural identity in favor of adopting the cultural identity of the host.

Separation desires to maintain all features of the heritage identity while rejecting relationships with members of the host culture.

Marginalization rejects both heritage and host cultures, losing meaningful contact with both.

Relevance and Implications of Acculturation Orientations

Navas and her colleagues (2005) remind us that the acculturation process is complex and differs according to acculturative contexts or domains. For instance, it is possible for teachers or students to select the integration strategy for school, the separation strategy for dating or mating, and the assimilation strategy for the workplace. In general, for both hosts and immigrants, separation and marginalization strategies govern ideological domains such as religion, values, or lifestyle, while integration and assimilation strategies to govern more material contexts such as work, holidays, foods, and dress (Navas et al., 2005, p. 29)

Strategies adopted by teachers can produce conflict and stress for immigrant students, creating social conflict for teachers, emotional exhaustion, and diversity-related burnout. Some stress is classified as problematic, meaning the host and immigrant will have some areas of agreement and some areas that will need negotiation and accommodation. Severe acculturative stress occurs when the two orientations have little agreement in any area. In that case, teachers and students must confront the fact that their situation will not be resolved quickly or easily. In the general society, such mismatched participants try to avoid each other; in classrooms, where avoidance is impossible, the best possibility is extensive negotiation and accommodation over a long period (Berry, 2005; Bourhis et al., 2009).

The Interactive Acculturation Model (Bourhis et al., 1997), displayed in Table 1, indicates the probable relational outcomes of host community and immigrant acculturation orientations. In a refinement of the Berry model, Bourhis and colleagues add another, more cosmopolitan orientation of "individualism," in which hosts or immigrants see themselves and others as individual persons rather than as members of any particular cultural or racial group.

Table 1. Probable Outcomes of Acculturation Orientation Interactions

Teacher Orientations	Immigrant Student Orientations				
	Integration	Assimilation	Separation	Marginalization	Individualism
Integration	Consensual	Problematic	Conflictual	Problematic	Problematic
Assimilation	Problematic	Consensual	Conflictual	Problematic	Problematic
Segregation	Conflictual	Conflictual	Conflictual	Conflictual	Conflitual
Exclusion	Conflictual	Conflictual	Conflictual	Conflictual	Conflictual
Individualism	Problematic	Problematic	Problematic	Problematic	Problematic

Source: Adapted from The Interactive Acculturation Model (IAM) by Bourhis et al., 1997, p. 382

As outlined in Table 1, the most harmonious interactions are typically between like-oriented teachers and students. Segregation and exclusion orientations encounter conflict with every orientation but their own. Teacher orientations of integration or assimilation have the most consensual interactions, while teachers with an individualist orientation have the fewest problems with social conflict (Bourhis et al., 1997). For example, when a teacher teaches for integration, she will include class activities that encourage students to share experiences and ideas from their culture. If the student aligns with this integration orientation, he will see these activities as opportunities to integrate. Conversely, when a teacher teaches for assimilation, the student who wants to be a separatist will reject lessons devoted to the values and beliefs of the dominant culture.

Acculturation Strategy Selection

Although acculturation orientation is usually an automatic reaction based on our psychology and previous social experiences, people can select and permanently change their orientations over time (Berry, 1997; Navas et al., 2005). For teachers encountering immigrant students, strategy selection can be critical. Internal and external factors influence individuals' preference of one orientation over another. Strategy selection can be based on many factors, such as gender, motivation, friendship ties, voluntariness, and the permanency of the move (Tadmor & Tetlock, 2006, p. 177). In addition, Reitz (2002, p. 1006) suggests four major dimensions of society that influence strategy selection:

- Pre-existing ethnic or race relations within the host population
- Differences in labor markets and related institutions
- The impact of government policies and programs, including immigration policy, policies for immigrant integration, and policies for the regulation of social institutions
- The changing nature of international boundaries, part of the process of globalization

For example, students who believe that their family is only visiting due to labor shortages in their home country may be armed with an acculturation strategy of separation, which puts them in conflictual relations with teachers selecting any but the individualism orientation. Imagine the stress of a teacher, beginning a school year, wrapped in her integration acculturation cloak, ready to welcome the diversity she hopes to find in the classroom, only to encounter a student, wrapped in her separation cloak, effectively announcing, "I don't want any part of your culture, and I'm only going to respond to your overtures toward my heritage culture." Anything short of the teacher shifting to an individualist strategy when dealing with this student may produce conflict with very stressful results for the entire classroom. Situations such as this one challenge teachers' power in the classroom, and as a result require a shift in identity to one of accommodator and collaborator in a shared enterprise, an educational project, as Freire refers to it, wherein all parties find liberation from cultural dominance or domination in a new structure, cooperatively built (Freire, 1970).

All of these factors combine to create a sense of accountability in beliefs, attitudes, and behaviors (Tadmor & Tetlock, 2006). Accountability in this case refers to "the need to justify one's thought and actions to . . . others, in accord with . . . shared norms" (p. 178). External pressures, such as "community economics, ethnic relations or biases, and federal and local immigrant policies, along with internal pressures, such as the internalized voice of someone with whom people feel strong affinity, shape the selection of acculturation strategies" (Tadmor & Tetlock, 2006; p. 179).

For instance, teachers may feel external accountability pressure from administrators and local citizens to keep yearly test scores high, according to the dictates of No Child Left Behind, which is stronger than the internal accountability pressure of nurturing immigrant students' heritage. In this case, teachers would be inclined to select an acculturation strategy of assimilation, even if their preferred orientation might be toward integration.

Navas and colleagues (2005) identified other factors as part of their model for acculturation, the Relative Acculturation Extended Model (RAEM). This model distinguishes between preferred and adopted acculturation strategies used by both hosts (teachers) and immigrants (students). Simply put, both teachers and students struggle between "ideal" and "real" strategy selection. To teachers, "ideal" means the strategy they would prefer to use; "real" means the strategy they select and use to adapt to the situation or the strategy used by students (Navas et al., 2005, p. 26). The example given earlier was of the teacher whose natural leaning was toward an integration orientation, but she felt compelled to adopt a strategy of assimilation, believing it might better her students' high-stakes test scores.

Further complicating the process of acculturation is the concept of transnationalism (TN). The hallmark of TN is the large number of contacts and communications with other countries (VanOudenhoven et al., 2006). TN students retain regular contact with their country of origin, consulting family and friends there for support and advice, minimizing their dependence upon classroom teachers. They transcend the ethnic hyphenates so prevalent in the United States, such as African-American, Mexican-American, or Asian-American, making TN students' identity naturally resistant to labeling and pigeonholing. Transnational students' identity is based on a concept of being from two or more places (across national borders) at once, as opposed to being from one nation, routinely moving back and forth between other nations.

Beyond transnationalism, research indicates that today's youth are exploring and developing a sense of global identity. In addition to navigating both their native culture and any other culture in which they are immersed or with which they have routine contact, modern technology has given adolescents' identities a wider scope, a "pan-human culture," that de-emphasizes focus on other cultural identities (Berry & Sabatier, 2011, p. 659). Both transnationalism and pan-humanism contribute to the orientation strategies that teachers must select in order to achieve an acculturational fit with those students.

Building a Classroom Culture

The common in-group identity model, as described by Gaertner and Dovidio (2005), allows members to retain their heritage culture while still embracing the goals of the aggregate, sliding between identities as the context requires. When teachers promote shared experiences for a feeling of "playing on the same team" (Gaertner & Dovidio, 2005, p. 630), they provide authentic opportunities for teachers and immigrant students to negotiate acculturation strategies. These negotiations facilitate interdependence and alter individual attitudes about the classroom culture.

Building a classroom culture of learners melds acculturation orientations through negotiation, communication, and continual monitoring of participation. Some considerations when establishing a classroom culture (CC) include the following:

- Evolve practices from "teacher as American culture transmitter" to "teacher as American culture transformer" (Cockrell et al., 1999). This process creates a cooperative, rather than assimilative, environment for learning while capitalizing on the best of all participants.
- Create ways to emphasize group collaboration over individual competition. Stressing cooperation and collaboration reduces marginalization in the classroom while promoting a sense of being a part of a larger group.
- Establish that "different" does not mean deficient or substandard; combining individual differences can create a new whole.
- Redirect stereotyping to individual actions and thoughts.
- Find ways to elicit knowledge that immigrants' prior school experiences can contribute to the collective. Luis Moll and colleagues remind us that immigrants have experiences and views that are valuable to building a better understanding of the world and our place in it (Moll, Amanti, Neff, & Gonzalez, 1992).
- Assign tasks that require group discussion to small groups and encourage use of hybrid language creations to include as many voices as possible.
- Allow sufficient time for negotiated collaboration of small groups for increased participation and creative, often hybrid, responses.
- Watch individual performance to ascertain strengths that can be publicly acknowledged to help equalize status within groups (Cohen & Lotan, 1995).

- Live a democratic ideal by negotiating with CC on how the classroom will operate and how it will resolve misunderstandings and disagreements.
- Teach from concepts and pictures (e.g., war, colonizing, poverty, immigration), rather than from discrete facts, to elicit responses from more students, creating a view of philosophies and understandings that reflect the CC. Add facts as understandings grow.
- Arrange short field trips to build common experiences for discussion and learning.

Understanding the acculturation process and seeking strategies to improve classroom culture serve to create a positive atmosphere, centered on diversity for teachers and students alike. In this way, teachers on the front line of societal change "act as both institutional agents, as well as agents of reception within the host society" (Dabach, 2011, p. 68), and guide the entire classroom in an acculturative adventure.

Conclusion

Globalization networks promise increased contact and communication among diverse peoples, worldwide, rivaling the development of the printing press for the proliferation of new ideas and opportunities. Along with the hopeful and exciting prospects come challenges inherent to any endeavor that involves interactions between individuals and groups. As families emigrate to take advantage of possibilities outside their homelands, they bring their children to be schooled within a new culture, putting teachers on the front line of global migration. While the new residents struggle with adjusting to their unfamiliar life, teachers, too, make adjustments in their own identities and coping strategies. The acculturative process for both the host and immigrant participants benefits from careful nurturing of negotiated interactions. By taking stock of their own acculturation orientation and being ready to deliberately and mindfully alter it to mesh with the orientation strategies of individual students, teachers can reduce acculturative stress leading to diversity-related burnout, while creating a classroom environment that helps facilitate the reception, integration, and academic success of their newest class member.

References

Batalova, J., & Terrazas, A. (2010). *Frequently requested statistics on immigrants and immigration in the United States.* Retrieved from http://www. migrationpolicy.org/ pubs/ integration-Jimenez.pdf

Berry, J. W. (1974). Psychological aspects of cultural pluralism: Unity and identity reconsidered. *Topics in Culture Learning, 2,* 17–22.

Berry, J. W. (1997). Immigration, acculturation, and adaptation. *Applied Psychology: An International Review, 46*(1), 5–34.

Berry, J. W. (2005). Acculturation: Living successfully in two cultures. *International Journal of Intercultural Relations, 29,* 697–706.

Berry, J. W. (2008). Globalization and acculturation. *International Journal of Intercultural Relations, 32,* 328–336.

Berry, J. W., & Sabatier, C. (2011). Variations in the assessment of acculturation attitudes: Their relationships with psychological wellbeing. *International Journal of Intercultural Relations, 35*(5), 658–669.

Bourhis, R. Y., Moise, L. C., Perreault, S., & Senecal, S. (1997). Towards an interactive acculturation model: A social psychological approach. *International Journal of Psychology, 32*(6), 369–386.

Chang, M. (2009). An appraisal perspective of teacher burnout: Examining the emotional work of teachers. *Educational Psychology Review, 21*(3), 193–218.

Cockrell, K. S., Placier, P. L., Cockrell, D. H., & Middleton, J. N. (1999). Coming to terms with "diversity" and "multiculturalism" in teacher education: Learning about our students, changing our practice. *Teaching and Teacher Education, 15,* 351–366.

Cohen, E. G., & Lotan, R. A. (1995). Producing equal-status interaction in the heterogeneous classroom. *American Educational Research Journal, 32*(1), 99–120.

Dabach, D. B. (2011). Teachers as agents of reception: An analysis of teacher preference for immigrant-origin second language learners. *The New Educator, 7*(1), 66–86.

Freire, P. (1970). *Pedagogy of the oppressed.* New York: Continuum.

Gaertner, S. L., & Dovidio, J. F. (2005). Understanding and addressing contemporary racism: From aversive racism to the common in-group identity model. *Journal of Social Issues, 61*(3), 615–639.

Horenczyk, G. (1997). Immigrants' perceptions of host attitudes and their reconstruction of cultural groups. *Applied Psychology, 46*(1), 34–38.

Horenczyk, G., & Tatar, M. (2002). Teachers' attitudes toward multiculturalism and their perceptions of the school organizational culture. *Teaching and Teacher Education, 18,* 435–445.

Kamp, A., & Mansouri, F. (2010). Constructing inclusive education in a neo-liberal context: Promoting inclusion of Arab-Australian students in an Australian context. *British Educational Research Journal, 36*(5), 733–744.

Kim, Y. Y., & McKay-Semmler, K. (2012). Social engagement and cross-cultural adaptation: An examination of direct- and mediated interpersonal communication activities of educated non-natives in the United States. *International Journal of Intercultural Relations.* Retrieved from http:// dx.doi.org/10.1016/j.ijintrel.2012.04.015

Maslach, C. (2003). Job burnout: New directions in research and intervention. *Current Directions in Psychological Science, 12*(5), 189–192.

Maslach, C. (2011). Burnout and engagement in the workplace: New perspectives. *European Health Psychologist, 13*(3), 44–47. Retrieved From http: //www.ehps.net/ehp/issues/2011/v13iss3_September2011/ EHP_ September_2011.pdf#page=10.

Migration Policy Institute (2012). *Immigrant children, urban schools, and the No Child Left Behind Act.* Retrieved from http://www.migrationinformation.org/datahub/countrydata.cfm?ID=347.

Moll, L. C., Amanti, C., Neff, D., & Gonzalez, N. (1992). Funds of knowledge for teaching: Using a qualitative approach to connect homes and classrooms. *Theory into Practice,* 31(2), 132–141.

Navas, M., Garcia, M. C., Sanchez, J., Rojas, A. J., Pumares, P., & Fernandez, J. S. (2005). Relative acculturation extended model (RAEM): New contributions with regard to the study of acculturation. *International Journal of Intercultural Relations,* 29(1), 21–37.

Phinney, J. S., Horenczyk, G., Liebkind, K., & Vedder, P. (2001). Ethnic identity, immigration, and well-being: An interactional perspective. *Journal of Social Issues,* 57(3), 493–510.

Reitz, J. G. (2002). Host societies and the reception of immigrants: Research themes, emerging theories and methodological issues. *International Migration Review,* 36(4), 1005–1019.

Sam, D. L., & Berry, J. W. (2010). Acculturation: When individuals and groups of different cultural backgrounds meet. *Perspectives on Psychological Science,* 5(4), 472–481.

Tadmor, C. T., & Tetlock, P. E. (2006). Biculturalism: A model of the effects of second-culture exposure on acculturation and integrative complexity. *Journal of Cross-Cultural Psychology,* 37(2), 173–190.

Tatar, M., & Horenczyk, G. (2003). Diversity-related burnout among teachers. *Teaching and Teacher Education,* 19, 397–408.

VanOudenhoven, J. P., & Hofstra, J. (2006). Personal reactions to "strange" situations: Attachment styles and acculturation attitudes of immigrants and majority members. *International Journal of Intercultural Relations,* 30(6), 784–798.

VanOudenhoven, J. P., Ward, C., and & Masgoret, A. M. (2006). Patterns of relations between immigrants and host societies. *International Journal of Intercultural Relations,* 30(6), 637–651.

Ward, C. (2001). The A, B, Cs of acculturation. In D. Matsumoto (Ed.), *The handbook of culture and psychology* (pp. 411–445). Oxford, England: Oxford University Press.

Ward, C. (2008). Thinking outside the Berry boxes: New perspectives on identity, acculturation and intercultural relations. *International Journal of Intercultural Relations,* 32(2), 105–114.

CHAPTER THREE

(En)countering the Paradox: Challenging the Neoliberal Immigrant Identity

Paul G. Fitchett

> Neoliberalism is an economic, political, and social doctrine emphasizing free-market values and placing private industry over public welfare. The immigrant community has emerged as a paradoxical topic for neoliberalism. Either valorized as nation-builders and sources of labor or minimized and castigated as a threat to Eurocentric traditions, the influx of immigrants across U.S. boundaries is a contentious side effect of this hyper-capitalist derivative of globalization. This chapter explores the dynamics of neoliberalism and its impact on the immigrant identity, and how social studies textbooks and curricula have propagated a neoliberal message toward immigration. I offer strategies for challenging the neoliberal immigrant identity in social studies classrooms.

According to Merryfield and Kasai (2004), globalization is the dynamic interconnectedness of culture, economies, and communities across nations. Under such pretense, globalization serves to ameliorate societal conflicts, foster interdependence, and recognize multiple ways of knowing (Spring, 2008). Yet, the looseness of defining globalization has also given way to another form, neoliberalism. Highly controversial and complex, neoliberalism surfaced as an economic theory placing emphasis on free-market ideology and a deregulation of the nation-state (Chomsky, 1999; Harvey, 2007). In its current iteration, neoliberalism is championed as an uninhibited capitalist approach toward international commerce and trade. On the ground, however, it has contributed to unstable geopolitical climates, the suppression of labor at the expense of capital, an assault on collective bargaining, and a dehumanizing of the workforce.

Under neoliberal doctrine, immigration patterns fluctuate as people search for more secure and less oppressive economic opportunities. From the 1980s to post-NAFTA 2004, the number of undocumented immigrants arriving in the United States increased from 140,000 a year to 700,000 (Passel, 2005). The immigrant community, both in a historical and contemporary context, has emerged as a paradoxical topic for neoliberalism. Either valorized as nation-builders and sources of labor or minimized and castigated as a threat to Eurocentric traditions, the influx of immigration across U.S. boundaries is a contentious side effect of this hyper-capitalist derivative of globalization (Varsanyi, 2009). This chapter explores the dynamics of neoliberalism, its impact on the immigrant identity, and how social studies textbooks and curricula have propagated a neoliberal message toward immigration. As I identify this neoliberal message, I also offer strategies for challenging its agenda in social studies classrooms.

Neoliberalism, Globalization, and Education

Chomsky (1999) defines neoliberalism as a political, economic, and/or social mentality that emphasizes free-market (choice) initiatives, private industry over public welfare, and profit as a measure of success. An extension of liberal theory as perpetuated by Adam Smith, neoliberalism developed into a dominant economic theory of the United States and Great Britain and, by extension, Western culture in the late twentieth century (Harvey, 2007). Signifying the collapse of the Soviet Union as evidence of socialism's failure, neoliberal political leaders have since promoted free-market entrepreneurship as a model of efficiency and progress, whereas state-controlled bureaucracies are perceived as inept and wasteful. The role of government under this unrestrained capitalist ideology is simply to protect business interests at home and overseas (i.e., regulate trade, provide defense, and maintain the infrastructure).

Neoliberalism, masked as globalization, often morphs into a disingenuous "global meliorism" (Chomsky, 1999, p. 92) by which corporations of leading nation-states set up shop in less developed nation-states in order to exploit cheap labor and secure access to natural resources. Protected by a government-sponsored military presence and a message of democracy-building, corporate financiers and captains of industry make promises of opportunity and freedom to local communities while simultaneously pursuing increasing profit. All human interaction and ethics are situated under the domain of the market. As a result, people and cultures are atomized and their relevance to society is determined by their usefulness as labor.

Neoliberal doctrine penetrates social institutions as well. Education, in particular, garners the attention of hyper-capitalist ideologues. School-choice programs, vouchers, and charters, educational archetypes of free-market thinking, have received substantial corporate backing (Parker & Camacia, 2009; Sleeter, 2012). For example, the Broad Foundations, founded by businessperson Eli Broad and his wife, Edythe, have invested hundreds of millions of dollars in education initiatives in the name of global competitiveness and entrepreneurship (Broad Foundations, 2012). Their far-reaching influence has provided scholarships and prize money to establish charter schools and produce business-oriented school administrators.

By extension, global and multicultural curriculum development and (de)emphasis are also influenced by neoliberalism. In many schools across the United States, international education is synonymous with career readiness and international competition (Camacia & Franklin, 2010; Parker & Camacia, 2009). Learners' exposure to the "other" is limited to labor exploitation or adversarial us-versus-them perspectives. The curriculum embedded in such schools implicitly discourages critical multicultural education in favor of superficial, tokenized multiculturalism (Sleeter, 2012) or denies the existence of racism in favor of a one-world, color-blind ideology (Macedo & Gounari, 2006; Myers, 2010). Racism does not align with the neoliberal ethos. Cultural bias and social hegemony conflict with the not-so-subtle push toward meritocracy and individual entrepreneurship. Discourse revolving around race, ethnicity, and difference is played down in favor of commonality. Pluralism is perceived as divisive. A two-tiered neoliberal message emerges: the free-market is always preferable and individuality trumps the collective. In the subsequent section, I analyze the identity of immigrants through this paradigm.

The Immigrant Identity: A Neoliberal Paradox

The neoliberal message that trickles down into social institutions and ideology propagates a sameness and one-world mentality at the expense of multiculturalism and diversity. In this narrow view, immigrants and their communal behavior represent a closed society in opposition to a rugged individualism ethic promoted in many European American cultures. Neoliberals censor any message that supports differences over sameness or critiques the majority for their subjugation over the minority (Macedo & Gounari, 2006; McLaren, 2000; Soto & Joseph, 2010).

Neoliberalism commoditizes the immigrant identity. As McLaren (2000) stated, "The market value of an immigrant's labor is increasingly becoming a central concern in the debates over multiculturalism" (p. 6). Immigrant contributions to the social fabric are minimized to superficial cultural celebrations/holidays and hero name-dropping (Sleeter, 2012). Thus, while immigrants are recognized (colonized) as a source of capital (Willinsky, 1998), they remain the "other" whose cultural contribution to the democratic nation-state is associated with the ubiquitous melting pot analogy. The contradictory market place and societal interests of neoliberalism further confound the identity of the immigrant. While promoting a borderless, unfettered capitalism, neoliberal advocates push for tighter immigration regulation, particularly against immigrant groups viewed as less economically self-sufficient and more culturally independent of the majority culture (Levine, Lopez, & Marcelo,

2008; Ong, 1996). Varsanyi (2007, 2009) refers to this tension as a neoliberal paradox in which immigrants are situated as a perceived threat to the dominant culture's conception of citizen while simultaneously exploited for cheap labor (Chomsky, 1999; Hayduk, 2009; Sánchez, 2011). Perpetuating the "push and pull" of immigration, market-driven policies encourage businesses to take advantage of less-regulated labor environments in developing nations and create a hostile living environment for the indigenous working class. Seeking a more advantageous lifestyle in which wages are higher and government is more stable, the workforce immigrates to developed nations where they find employment as cheap, interchangeable labor. Socially, this proliferation of immigration is often met with xenophobia, racism, and fears of cultural encroachment (Mohl, 2009). Thus, neoliberalism defines the immigrants' identity by this paradox: they are a low-cost workforce on hand (economic benefit) while simultaneously they are perceived as culturally insulated and a financial drain on the nation-state (societal detriment).

Perhaps no other trade agreement and its effects exemplify the neoliberal paradox as does the North American Free Trade Agreement (NAFTA), a trade policy providing relative free trade between Canada, the United States, and Mexico (North American Free Trade Agreement, 1993). Enacted in 1994, it received bipartisan support not only for its desired goal to free up U.S. markets, but also to decrease the number of undocumented immigrants arriving into the United States from Mexico (Sánchez, 2011). However, NAFTA policies allowed U.S. industries and agribusinesses to relocate to Mexico where they outcompeted local industry and farmers, thereby increasing unemployment in many economic sectors. Unable to find work in their own country, Mexican families migrated to the United States in search of better job opportunities. Consequently, the number of undocumented immigrants in the United States rose by 300,000 people (from 450,000 to 750,000) in the first four years following the inception of NAFTA, outpacing documented immigrants (Passel, 2005). The new influx of immigrants proved a cheap source of labor for many U.S.-housed businesses while simultaneously displacing American workers. This shift in the labor pool magnified racism, resentment, and anti-immigrant vitriol within many U.S. communities, which brand these immigrants as criminals and job stealers (Mohl, 2009; Sánchez, 2011). In contradictory fashion, undocumented immigrants bolstered U.S. business profits through their low-wage employment while simultaneously being castigated for their non-native culture and perceived drain on the welfare of the state.

Social Studies Curriculum and Textbook Promotion of Neoliberalism

Curricula and textbooks, as canons for social reproduction and cultural hegemony (Apple, 2004; Apple & Christian-Smith, 1991; Aronowitz & Giroux, 1991; Bordieu & Passeron, 1990), frequently perpetuate the two-tiered neoliberal message and its paradoxical stance on immigration. The communal, as opposed to individual, identity of immigrant groups is a perceived threat to the individualistic, "Freedom Quest" message (McLaren, 2000; Spring, 2009; VanSledright, 2011). American schooling, in particular social studies, frequently downplays the pluribus in favor of the unum. Curricula identify immigrants both as an impediment to the development of a unified American identity and as an important cog in the nation's industrial machine, in yet another turn of the neoliberal paradox.

State history curricula portray immigrants as subversives or as a labor class important in the building of America (Fitchett & Salas, 2010; Journell, 2009). Narratives that do not mold into this image are denied access to the curriculum. Frequently missing from the canon are cultural contributions, social movements, and worker exploitation. McLaren (2000, p. 4) refers to this phenomenon as a "historical amnesia," in which capitalism's violent past and present are forgotten in order to preserve the sanctity of the neoliberal ideology. Thus, the important contribution of the Bracero movement in the 1940s, the social justice work of César Chávez, and the appalling fate of Chinese laborers on the Transcontinental Railroad remain footnotes in the narrative.

While neoliberalism advocates push for school privatization, they also seek to use state and federal governments to craft a curriculum advantageous to their message or deny access to those curricula deemed contrarian (Spring, 2008). For example, consider the decision by the Arizona state legislature to eliminate ethnic studies from the public school curriculum. State school board officials, the state superintendent, and the governor of Arizona claimed ethnic studies courses propagate anti-American rhetoric and support radicalism (Lewin, 2010). In that contemporary hotbed for anti-immigrant fervor, lawmakers based their decision, ostensibly, on a fear that critical discourse on diversity would contribute to resentment and hatred. Soto and Joseph (2010, p. 50) suggest that the law represents a consistent refrain of neoliberal identity politics: "Any attention to race or racism, even as a topic of study, is itself racist." As such, a curriculum that emphasizes issues of diversity is a barrier to the social uniformity required for neoliberal policies to flourish. Rather, schools should promote individual rather than cultural identity for the sake of developing a complient, homogenous workforce (Macedo & Gounari, 2006). This places the non-white "other" in a precarious curricular position. Their collective identity and curriculum prioritization are framed within the context of neoliberalism.

As the most referenced cultural artifacts and curricular tools (Hilburn & Fitchett, 2012; Sleeter & Grant, 1991), textbooks tend to reflect a similar neoliberal emphasis of immigrant peoples as labor. The narratives often typecast the immigrant experience as "rags to riches" and "land of opportunity" metaphors contingent upon immigrants' hard work and usefulness in the building of the nation (Loewen, 1995, p. 209) or valorize immigrants for nation-building (Sleeter & Grant, 1991). Examples of racism, bigotry, and violence are subtly glossed over as incidental and unfortunate. Immigrants' contributions to the nation's cultural democracy are minimized, included only when situated within the dominant culture. Analogous to the neoliberal paradox, textbooks reference current immigrants' entrepreneurial spirit and hard-work ethic while simultaneously labeling immigrants as educationally deficient and socially burdensome (Hilburn & Fitchett, 2012).

Thus, textbooks and the formal curriculum maintain a socio-cultural status quo and promote dominant ideologies (Apple, 2004; Aronowitz & Giroux, 1991). Neoliberalism, as one such ideology, stigmatizes the historical positionality of immigrants. Consequently, immigrant students run the risk of disconnecting with the majority culture (Salinas, 2006). For members of the majority culture, the implicit curricular and textbook messages validate skepticism as to immigrant behavior and motivations. In the final section, I propose a model for working against the neoliberal message in order to foster a discourse that respects difference, instills a healthy critique of social and government institutions, and encourages a humanizing globalization.

Focusing on Community and Discourse to Counteract Neoliberal Stigmatization

As educators, we are often preoccupied with how we are going to engage the curriculum. Globalization under the neoliberal guise views the twenty-first-century world outside the United States as one gigantic and adversarial marketplace (Friedman, 2007). It supports a technocratic curriculum, perhaps best exemplified by the Common Core State Standards Initiative's (2010) mission statement:

> The standards are designed to be robust and relevant to the real world, reflecting the knowledge and skills that our young people need for success in college and careers.

With American students fully prepared for the future, our communities will be best positioned to compete successfully in the global economy.

Economics and job creation overshadow human interaction. The immigrant, as portrayed in such a curriculum, is at best minimized as labor and at worst demonized for cultural exclusivity. In this final section, I offer a theoretical framework for rethinking the socio-historical place of immigrants based on two key principles: building a curriculum of community and promoting democratic discourse.

Building a Curriculum of Community

Counteracting the neoliberal immigrant identity is a move beyond the pejorative, formalized curriculum toward a curriculum of community. The formal curriculum, typically prescribed and essentialized strands of Eurocentric content (Cornbleth & Waugh, 1995), communicates a deleterious message to both immigrant and native students that the "other" is unimportant or subsidiary within the master historical narrative (Salinas, 2006). Educators wanting to challenge the neoliberal message of individualism over commonality should build a curriculum of the community. By community, I do not refer to a national collective memory, but a community of plurality and multiple identity (Camacia & Franklin, 2010; Feinberg, 1998). This requires recognizing the cultural and social contributions of immigrant children (Moll, Amanti, Neff, & González, 1992). Including the culture of the home and embedding it within the sphere of the public curriculum necessitates a deep understanding of one's students and where they coexist within the curriculum. Immigrant students often maintain transnational ties, which influence their identity in a nation-state (Rong & Priessle, 2009). The neoliberal ideologue dismisses this aptitude as evidence of a deficient closed society that drains social services and avoids income taxes. Educators should instead embrace this cross-cultural exchange as an example of immigrants' "funds of knowledge" that enables them to move freely between societies (Moll et al., 1992; Moll & Arnot-Hopffer, 2005). Building community by nurturing cultural inclusiveness encourages students to identify with the formal curriculum in a meaningful way.

Rather than dismissing immigrant students as deficient in skills and knowledge, community-building teachers recognize the rich experiences of these children for their teaching and learning opportunities. Consider a Guatemalan student whose parents witnessed firsthand the oppression of United Fruit Company and took part in organized political resistance against the corporation's practices. They offer a contemporary example of how laborers can turn to political parties and unions in order to combat the corporate-backed government regimes. The community-building teacher taps into this fund of knowledge, using (not exploiting) this experience by asking students and their parents to share their personal histories to teach the class the importance of political and social engagement as a lever for change.

In addition, the push and pull dynamics of global economics combined with changing immigration laws have contributed to a proliferation of undocumented immigrants arriving, for the most part, from developing nations. Paradoxical neoliberal curricular and textbook messages highlight the importance of these individuals as cheap labor while simultaneously casting them aside as a blight on the welfare system. Deconstruction and critical analysis of this paradox provide an educational opportunity for educators who seek to embrace a community-oriented pedagogy. Marri (2005) suggests beginning with a curriculum-questioning framework: Who is included and how are they represented? For teachers this involves examining the classroom demographics and reflecting on how students are included and represented in the formal curriculum. For example, how often are Mexican Americans mentioned in the U.S. history curriculum? How are Mexican Americans characterized in the curriculum? As pointed out earlier, the portrayal of the immigrants is full of contradictions and stigmatization (Hilburn & Fitchett, 2012; Journell, 2009). The community-minded teacher challenges students to explore the "why" of immigrant issues, exposing the violent and exploitative nature of unregulated free-market global economics and its societal impact.

Building a curriculum of community requires both the recognition of immigrants' knowledge as a legitimate source and a healthy critique of the traditional narrative. Moving beyond static representations of immigrants involves a willingness to enact critical analysis of the curriculum (Ross, 2006) and propagate a cultural democratic message (Banks, 2008; Parker, 2003). In the next step, I describe how promoting cultural democracy, via discourse, supports a curriculum of community.

Promoting Democratic Discourse

For teachers to counter the neoliberal curricular message toward immigrants and foster community, they should also engage students in a humanizing pedagogy (Bartolomé, 1994). The first step in this pedagogical process is dismissing the one-world, end-of-racism rhetoric. As an alternative, educators need to recognize how racialized national institutions and policies influence both the experience and portrayal of non-white groups (including many immigrant groups) in this country (Chandler, 2010; Ladson-Billings, 2003). Teachers promoting a humanizing pedagogy distinguish multiple cultural realities and advocate no single cultural position (Spring, 2008). As a pedagogical tool for making sense of these different cultural positions, democratic discourse provides students the opportunity to discuss the complex immigrant identity.

Freire (1998) suggests that neoliberal education can be subverted by embracing learners' position within the curricular landscape and fostering a dialectic praxis among learners and teacher. In other words, giving value to the voice of immigrant students and respecting their personal narratives, rather than stigmatizing them, is a powerful tool. Previous research of democratic discourse (Hess, 2009; Parker, 2003) indicates that creating space for discussion is an appropriate pedagogy for mediating conflicting values. In an earlier paper (Fitchett, Portes, & Salas, 2010), my colleagues and I offered an epistemological scaffold for encouraging discourse on the issue of immigration. First, we argue that teachers have to present several sides to the immigration issue, a complexity that does not align with the contradictory economic focus of a neoliberal global agenda. Students should examine both the motivations for immigrant arrival to the United States (both economic and political) and the contributions that immigrants offer (both economic and cultural) within our society. Developing into critical consumers of the immigrant curriculum (Ross, 2006), students then break away from banal textbooks and engage in inquiry-based instruction of documents and other historical artifacts related to immigration. Analysis of alternative source material provides learners the opportunity to engage in immigrant issues from a nonessentialist position (VanSledright, 2004). As a final step, students take part in structured, democratic discussion for the purpose of mediating the competing arguments surrounding immigration. When students engage in deliberative discourse, they are challenged to examine aspects of an argument or position contrary to their own (Parker, 2003). Instead of engaging in a winner-takes-all rant, participants find a middle ground (consensus) from which to build toward an enlightened understanding. Using deliberative discourse models, such as structured academic controversy models (Johnson & Johnson, 1988), learners come to understand that the implicit market-driven neoliberal message of immigrants is not transparent but instead muddied by the complexities of economic exploitation and cultural exchange. Thus, systemic issues of race, place, and marketplace values shape the curricular position of the immigrant.

Conclusion

The teaching of complex, controversial issues of globalization is a powerful tool for preparing learners in their role as cultural democratic participants

(Merryfield & Kasai, 2004). One such controversial (and highly complex) component of globalization is the growing dominance of neoliberal policy and ideology, specifically concerning how it shapes our view of immigrants. Neoliberalism as a socio-cultural ideology places market-driven values ahead of human rights. Immigrants, many of whom have been personally affected by neoliberal practices, are pigeonholed into two contradictory categories: economic laborer or social welfare leech. This position is further problematized by the paradoxical relationship of liberal economic practices and exploitation of labor, which have contributed to the most recent influx of immigrants. While textbooks and other forms of curricula attempt to essentialize the immigrant experience, I have suggested two principles for counteracting this neoliberal stigmatization. Building from community, educators value the private, home culture as a valuable asset in enacting the formal curriculum. They also encourage pedagogy for deliberating the competing values surrounding immigration and its place in contemporary America.

Understanding the role of immigration in the global-education landscape requires more than complacently accepting a one-world, noncontroversial dogma. The world is not flat, but a curved, complex entity that cannot (should not) be commoditized. Realizing a humanizing globalization requires critiquing the hidden neoliberal message and appreciating the multifaceted realities of our society.

References

Apple, M. W. (2004). *Ideology and curriculum*. New York: RoutledgeFalmer.
Apple, M. W., & Christian-Smith, L. K. (1991). The politics of the textbook. In M. W. Apple & L. K. Christian-Smith (Eds.), *The politics of the textbook* (pp. 1–23). New York: Routledge.
Aronowitz, S., & Giroux, H. A. (1991). Textual authority, culture, and politics of literacy. In M. W. Apple & L. K. Christian-Smith (Eds.), *The politics of the textbook* (pp. 213–241). New York: Routledge.
Banks, J. A. (2008). Diversity, group identity, and citizenship in a global age. *Educational Researcher, 37*(3), 129–139.
Bartolomé, L. I. (1994). Beyond the methods fetish: Toward a humanizing pedagogy. *Harvard Educational Review, 64*(2), 173–194.
Bordieu, P., & Passeron, J. C. (1990). *Reproduction in education, society, and culture* (2nd ed.). London: Sage.
Broad Foundations. (2012). *The Broad Foundation: Education*. Retrieved from http://www.broadeducation.org/index.html
Camacia, S. P., & Franklin, B. M. (2010). Curriculum reform in a globalised world: The discourses of cosmopolitanism and community. *London Review of Education, 8*(2), 93–104.
Chandler, P. (2010). Critical race theory and social studies: Centering the Native American experience. *Journal of Social Studies Research, 34*(1), 29–58.

Chomsky, N. (1999). *Profit over people: Neoliberalism and global order.* New York: Seven Stories Press.
Common Core State Standards Initiative. (2010). *Common Core State Standards Initiative: Mission Statement.*, Retrieved from http://www.corestandards.org/
Cornbleth, C., & Waugh, D. (1995). *The great speckled bird: Multicultural politics and education policymaking.* Mahwah, NJ: Erlbaum.
Feinberg, W. (1998). *Common schools/uncommon identity: National unity and cultural difference.* New Haven, CT: Yale University Press.
Fitchett, P. G., Portes, P., & Salas, S. (2010). "You lie—that's not true": Immigration and preserve teacher education. *Action in Teacher Education*, 32(4), 96-104.
Fitchett, P. G., & Salas, S. (2010). Latinos, counteraction, and the (hidden) social studies curriculum: A cultural-historical theory and praxis for teacher education. *Border-Lines*, 4, 39-60.
Freire, P. (1998). *Pedagogy of Freedom.* Oxford, England: Rowman & Littlefield.
Friedman, T. L. (2007). *The world is flat: A brief history of the twenty-first century.* New York: Macmillan.
Harvey, D. (2007). *A brief history of neoliberalism.* Oxford, England: Oxford University Press.
Hayduk, R. (2009). Radical responses to neoliberalism: Immigrant rights in the global era. *Dialectical Anthropology*, 3, 157-173.
Hess, D. E. (2009). *Controversy in the classroom: The democratic power of discussion.* New York: Routledge.
Hilburn, J., & Fitchett, P. G. (2012). The new gateway, an old paradox: Immigrants and involuntary Americans in North Carolina history textbooks. *Theory and Research in Social Education*, 40(1), 35-65.
Johnson, D. W., & Johnson, R. T. (1988). Critical thinking through structured controversy. *Educational Leadership*, 45(8), 58-64.
Journell, W. (2009). Setting out the (un)welcome mat: A portrayal of immigration in state standards for American history. *The Social Studies*, 100(4), 160-168.
Ladson-Billings, G. (2003). Lies my teacher still tells: Developing a critical race perspective toward the social studies. In G. Ladson-Billings (Ed.), *Critical race theory perspectives on the social studies* (pp. 1-14). Greenwich, CT: Information Age.
Levine, P., Lopez, M. H., & Marcelo, K. B. (2008). *Getting narrower at the base: The American curriculum after NCLB.* Medford, MA: Center for Information & Research on Civic Learning and Engagement.
Lewin, T. (2010). Citing individualism, Arizona tries to rein in ethnic studies in school. *The New York Times*, p. A13.
Loewen, J. W. (1995). *Lies my teacher told me: Everything your American history textbook got wrong.* New York: Simon & Schuster.
Macedo, D., & Gounari, P. (2006). Globalization and the unleashing of new racism: An introduction. In D. Macedo & P. Gounari (Eds.), *The globalization of racism* (pp. 1-36). Boulder, CO: Paradigm.
Marri, A. R. (2005). Building a framework for classroom-based multicultural democratic education: Learning from three skilled teachers. *Teachers College Record*, 107(5), 1035-1059.
McLaren, P. (2000). Democracy sabotaged by democracy: Immigration under neoliberalism. In E. T. Trueba & L. I. Bartolomé (Eds.), *Immigrant voices: In search of educational equity* (pp. 1-16). Oxford, England: Rowman & Littlefield.
Merryfield, M. M., & Kasai, M. (2004). How are teachers responding to globalization? *Social Education*, 68(5), 354-358.

Mohl, R. A. (2009). Globalization and Latin American immigration in Alabama. In M. E. Odem & E. Lacy (Eds.), *Latino immigrants and the transformation of the US south* (pp. 51-69). Athens: University of Georgia Press.

Moll, L. C., Amanti, C., Neff, D., & González, N. (1992). Funds of knowledge for teaching: Using a qualitative approach to connect homes and communities. *Theory into Practice*, 31(2), 132-140.

Moll, L. C., & Arnot-Hopffer, E. (2005). Sociocultural competence in teacher education. *Journal of Teacher Education*, 56(3), 242-247.

Myers, J. P. (2010). The curriculum of globalization. In B. Subedi (Ed.), *Critical global perspectives* (pp. 103-120). Charlotte, NC: Information Age.

North American Free Trade Agreement (NAFTA), Can.-Mex.-U.S., 32 I.L.M. 289 (1993).

Ong, A. (1996). Cultural citizenship as subject-making: Immigrants negotiate racial and cultural boundaries in the United States. *Current Anthropology*, 37(5), 737-751.

Parker, W. C. (2003). *Teaching democracy: Unity and diversity in public life*. New York: Teachers College Press.

Parker, W. C., & Camacia, S. P. (2009). Cognitive praxis in today's "international education" movement: Intents and affinities. *Theory and Research in Social Education*, 37(1), 42-74.

Passel, J. S. (2005). *Unauthorized migrants: Numbers and characteristics*. Washington, DC: Pew Hispanic Center.

Rong, X. L., & Priessle, J. (2009). *Educating immigrant students in the 21st century*. Thousand Oaks, CA: Corwin Press.

Ross, E. W. (2006). The struggle for the social studies curriculum. In E. W. Ross (Ed.), *The social studies curriculum: Purposes, problems, and possibilities* (3rd ed., pp. 17-36). Albany, NY: SUNY Press.

Salinas, C. (2006). Educating late arrival high school immigrant students: A call for a more democratic curriculum. *Multicultural Perspectives*, 8(1), 20-27.

Sánchez, H. E. (2011). Disposable workers: Immigrants after NAFTA and the nation's addiction to cheap labor. *Border-Lines*, 5, 44-68.

Sleeter, C. E. (2012). Confronting the marginalization of culturally responsive pedagogy. *Urban Education*, 1-23. doi:10.1177/0042085911431472

Sleeter, C. E., & Grant, C. E. (1991). Race, class, gender, and disability in current textbooks. In M. W. Apple & L. K. Christian-Smith (Eds.), *The politics of the textbook* (pp. 78-110). New York: Routledge.

Soto, S. K., & Joseph, M. (2010). Neoliberalism and the battle over ethnic studies in Arizona. *Thought & Action*, 45-56.

Spring, J. (2008). Research on globalization and education. *Review of educational research*, 78(2), 330-363.

Spring, J. (2009). *Deculturalization and the struggle for equality: A brief history of the education of dominated cultures in the United States* (6th ed.). New York: McGraw-Hill.

VanSledright, B. (2004). What does it mean to think historically and how do you teach it? *Social Education*, 68(3), 230-233.

VanSledright, B. (2011). *The challenge of rethinking history education: On practices, theories, and policy*. New York: Routledge.

Varsanyi, M. W. (2007). Documenting undocumented migrants: The matriculas consulares as neoliberal local membership. *Geopolitics*, 12, 299-319.

Varsanyi, M. W. (2009). *Rescaling the "alien," rescaling personhood: Neoliberalism, immigration, and the state*. San Diego, CA: Center for Comparative Immigration Studies.

Willinsky, J. (1998). *Learning to divide the world*. Minneapolis: University of Minnesota Press.

CHAPTER FOUR

Immigration and Global Economies in the Context of Globalization

Lydiah Nganga
Keonghee Tao Han

> In this chapter we examine immigration and global economies in the context of globalization. Globalization has led to new forms of colonialism and slavery, where powerful corporations from developed nations are taking over the land and natural resources of the Third World and forcing foreign-born immigrants to labor under inhumane conditions in host countries. In this neocolonial global context, we argue that global and multicultural education can be used to resolve global injustices and conflicts by helping instill in students a greater awareness of global interdependence and the need for social justice for all.

Introduction

In this chapter, we examine immigration issues related to economic globalization and how these issues call for K–12 teachers and university instructors to become well-informed and proactive global multicultural educators. We explore the influence of international immigration, labor migration, economic marginalization, and displacement of local people by multinational corporations. Our hope is that in the face of the current phenomenon of an economically driven global diaspora, multicultural education can offer a possible solution to human rights and social injustices.

Globalization and Its Worldwide Impact

Globalization refers to internationalization and interconnectedness of the well-being of nations, the increasingly interdependent nature of the businesses, peoples, and the world as a whole (Green & Griffith, 2002; Kambutu & Nganga, 2008). Economic globalization means that economies are interdependent and integrated on a worldwide basis. Economic globalization has resulted in global communication and technology, international trade and business governance and agreement, increasing corporate power, and social inequality (Green & Griffith, 2002), and worldwide migration and movements for economic reasons (Segal, Elliott, & Mayadas, 2010). Globalization has brought interconnections that transcend regional and national borders. As a result, issues and decisions made in one part of the world now have significant influence in far distant places (Akokpari, 2006). In some cases globalization has also led to increased levels of poverty. For example, Third World countries lose their land and resources to powerful multinational corporations that take over businesses and agriculture, while immigrant men and women in highly developed and developing nations are exploited without having recourse to the social rights and welfare benefits of their host countries.

Economic globalization has proved to be successful only in making global corporations and a few elites wildly wealthy. Contrary to the claims of globalization's proponents, wealth generated by globalization does not trickle down. Rather, the rules lock the wealth at the top, removing from governments and communities the very tools necessary to redistribute wealth and protect domestic industries, workers, social services, the environment, and sustainable livelihoods (Mander, Baker, & Korten, 2001). Consequently, globalization has not been limited to the movement of goods, technologies, and ideas but also applies to people moving across regional, national, and transitional borders. According to Gibson and Rojas (2006), immigration has become one of the undeniable by-products of globalization that is transforming the world socially, culturally, and politically. The number of the world's migrants (defined as persons living outside their country of birth) has doubled since 1960, reaching 191 million (United Nations Population Division [UNDP], 2006). Although economic globalization brought international trade and investment; market-driven technology; and economic, political, and cultural expansion, "globalization has led to benefits for some, but it has not led to benefits for all" (Anderson, 2002; Green & Griffith, 2002, p. 58). For example, women face racial and gender-based exploitation in their new countries (Kang, 2010). In addition, market-driven corporate dominance in the world and international division of wealth based on class, race, gender, national origin, ability, and religion benefited those who already have and further marginalized the have-nots in the integrated world economic system (Bhuyan, 2010; Choi, 2010; Green & Griffith, 2002).

The Interconnectedness of Globalization and Education

In this context, the world economic system is closely linked to education (Apple, 2010). Under such a premise, schooling serves the political ideology and cultural values of the mainstream, ruling-class group. We concur with Apple (2004, 2010) that economic globalization and education are interconnected. That is, school success is directly linked to economic gain and privilege in the job market, which in turn are connected to political power (Apple, 2004). The succession of the dominant group's power, privilege, and cultural values is preserved through the education system. Schools legitimize the dominant group's cultural forms, knowledge, and values while subjugating all others. In the global context, ruling powers of developed nations take over language use, cultures, and economies in less developed nations. Thus, "the asymmetries of power exist between nations and colonial and neocolonial histories, which see differential national effects of neoliberal globalization" (Lindgard, 2007, p. 239). Apple and colleagues (2010) and others (e.g., Green & Griffith, 2002) acknowledge gloomy realities of globalization and seek cultural pluralism through education. We agree with this position as a counter to "the relations of dominance and subordination" that are occurring in many countries at the present time (Apple, 2010, p. 3).

Immigration in the Absence of Comprehensive International Immigration Law

Several theories, albeit contradictory, have been proposed to explain why people migrate. The migration networks theory is one of the most prevalent. It explains the relationships between migrants and friends and family members at home who act as an information network (Haas, 2010; Massey et al., 1998). Those who immigrate first build a social capital that facilitates further migration. Another theory, the push-pull theory, posits that international migration is related to the global supply and demand for labor (Bauer & Zimmermann, 1998; Leighton Schwartz & Notini, 1994). Nations with scarce labor supply and high demand will have high wages that pull immigrants in from nations with a surplus of labor. One other theory that has been used to explain migration is the world systems theory, which suggests that international immigration is a by-product of global capitalism (Mabogunje, 1970). Whatever the reasons for immigration, the International Organization for Migration ("Illegal acquisitions," 2010) reported that the number of international migrants was estimated at 214 million in 2010.

Migration literature has highlighted in particular the key role of economic globalization in migration systems and migration networks (Haas, 2010). Notwithstanding these realities, cross-border movement of people and labor remains tightly controlled due to restrictive immigration laws and policies that have been put in place to uphold the principle of state sovereignty. In spite of these restrictions, the numbers of migrants are projected to increase in the twenty-first century (Anderson, 2010). Most global migrants are migrant workers. These are people who migrate for employment and for reunification with their families. International migration is therefore more of "a decent work and labor market issue than (an issue of) asylum-seekers or refugees with (security) issues" (Wickramasekara, 2008, p. 1249). The former UN Secretary General Kofi Annan stated that "the vast majority of migrants are industrious, courageous, and determined. They do not want a free ride. They want a fair opportunity. They are not criminals or terrorists. They are law abiding. They don't want to live apart. They want to integrate, while retaining their identity" (quoted in Wickramasekara, 2008, p. 1249).

To protect the rights of immigrants, the Universal Declaration of Human Rights recognizes that people have the right to move to any country and from any country, including their own. Despite this reality, there are no international laws that provide the right to enter to stay or work in a third country, since this would mean surrendering such rights to international treaties. Subsequently, restrictive immigration policies and a backlash from developed economies have revealed that immigrants are seen as taking over the jobs of local workers. This phenomenon has resulted in bias and stereotyping of immigrants, as well as discriminative practices and work conditions consistent with the need for them to take those jobs that the native citizens disdain and so avoid.

Theoretically, globalization establishes a global village where there should be unrestricted movement. In reality, there are practical barriers to the mobility of people across national borders. Such barriers are put in place by the receiving countries fearful of being overwhelmed by an influx of immigrants. Nevertheless, the need for cheap labor and raw materials by developed nations has led to trade negotiations between developed nations and developing countries. Developing and developed nations working in concert have created friendly international trade pacts such as the 1994 North American Free Trade Agreement (NAFTA). NAFTA opened Mexico to U.S. corporations, which flooded the Mexican market with underpriced products from the U.S. agricultural industry. As a result,

> small farms and small and medium-sized firms in Mexico went under, sending waves of displaced workers and families to Mexican cities and to the U.S., where "low skill" jobs were available in construction, in the growing service industry, and in the remnants of U.S. agricultural and manufacturing sectors. Similar processes of economic restructuring and displacement have occurred throughout the world and especially in Latin American countries such as El Salvador and Guatemala. (Apple, 2010, p. 50)

Following NAFTA, the United States signed other trade agreements with Peru, Jordan, and Chile, as well as leading the creation of the Central American

Free Trade Agreement (Bacon, 2008). These trade agreements had effects similar to those of NAFTA on local communities. Notwithstanding, debates over immigration policies continued in the United States "as though those trade agreements bore no relationships to the waves of displaced people migrating to the United States, looking for work" (Bacon, 2008, p. 23). Massive inflows of people are still entering the U.S. borders (il)legally as of today.

In other parts of the world, European and U.S. trade policies have negatively affected local populations by slashing the prices of locally produced goods for the benefit of developed nations, damaging the economies of Third World countries, and marginalizing local populations (Don & Gretchen, 1991). In Kenya and in Latin American countries such as El Salvador, Colombia, Ecuador, and Guatemala, multinational corporations have also displaced local people, creating large plantations to generate raw materials for developed nationals. Although these corporations provide jobs for the local people, most of those jobs are associated with precarious working conditions (Korovkin & Sanmiguel-Valderrama, 2007)

Immigration and Unfair Treatment of Migrants in Europe, Africa, the Middle East, and Asia

Europe. Immigrant groups moving to France before 1982 were mainly of European descent (from Italy, Spain, and Portugal), but since then immigrants from African countries such as Algeria, Morocco, Tunisia, Cameroon, Congo, Ivory Coast, Senegal, and Mali have become predominant (Michalowski, 2010). Other, smaller immigrant groups have arrived from Turkey and China, among other nations. Although immigration policies in France are favorable and guarantee access to full citizenship to children and grandchildren of former immigrants, it is still not an equal society. Migrants and their families often find themselves facing overt discrimination and prejudice. In other European nations, immigrants are faced with similar predicaments. Zufiaurre (2006), for example, identified immigration problems in Spain that have existed since the mid-1990s. Spain had been an exporter of migrants to other European nations as well as to the United States. However, that trend has now changed and Spain is experiencing a major wave of immigrants. Arriving from South America, Eastern Europe, and North Africa, immigrants are considered necessary for economic development yet are viewed with fear, a reaction that is common in other European countries.

The United Kingdom has received immigrants for centuries. However, the U.K. has seen increased numbers of immigrants in recent years (Somerville &

Cooper, 2010). This sudden increase is partially the result of increased numbers of international students, asylum seekers, family reunifications, and economic migration. Nonetheless, the immigrant population in the U.K. is very diverse, coming from Asian countries such as India, Pakistan, and China and from African countries such as Kenya, Ghana, Nigeria, and Uganda. Other immigrants come from Poland and the Middle East. The U.K. used a multicultural migration model for a long time but has since moved to an integration model. This means that the U.K. implements policies that "deliberately keep people marginalized and excluded through policies, controls, and differential support," thus discrimination exists (Elliott, Segal, & Mayadas, 2010, p. 453).

In Germany, the integration of immigrants is determined by their residential status. This also determines if immigrants have access to the labor market, social security, and other opportunities guaranteed to German citizens. Thus, immigrants of German descent are given preference. Other challenges faced by immigrants include, but are not limited to, acts of violence (Schmelz, 2010).

Africa. The historical background of immigration in Africa is indicative of the dynamics of cross-border migration. Migrants from less-developed African nations move to the more developed parts of Africa, seeking economic opportunities. In the early nineteenth century, there was a wave of migrants entering South Africa to work in the gold and diamond mines (Trimikliniotis, Gordon, & Zondo, 2008). Other countries that have experienced waves of immigrants are Kenya, Nigeria, Tanzania, Libya, and Ghana. Immigrant issues are similar to those in other nations.

Egypt has been an immigrant nation for thousands of years. Immigrants originate from Palestine, Sudan, Somalia, Libya, and other African countries (mostly as refugees). As is the case in other countries, voluntary migrants who are highly skilled international workers enjoy decent living conditions in Egypt. However, unskilled refugees and other foreigners experience discrimination in their daily lives. Native Egyptians shun immigrants and blame them for taking jobs and resources (Zohry, 2010). Kenya has also had an influx of immigrants from Somalia, Ethiopia, and the Sudan ("Illegal Acquisition of Citizenship," 2010), causing concerns over illegal immigration and acquisition of citizenship by Somalis.

The Middle East. In Israel, new immigrants have arrived in massive numbers since the 1970s. The Law of Return allowed migrants and family members to enter the country and receive citizenship promptly upon arrival. Jews of European and American descent, Jews of Asian and African origins, and other groups such as Palestinians differ in educational achievement and represent a broad socio-economic spectrum (Rajiman & Kemp, 2010). Non-Jewish foreign workers and Ethiopian Jews are placed at the margins of the Israeli economy, and suffer from the effects of racial stratification and discrimination.

Israel expresses restrictive conditions for immigrants similar to that of European nations. However, Israel presents a complex xenophobic and ethnonational prejudice against ethnically and religiously different non-Jewish groups, particularly against Palestinians. Palestinians living in Israel's governmental system have been subject to cultural and economic hegemony (Meshulam & Apple, 2010). Israeli policies and practices for non-Jews are highly restrictive. Although counterhegemonic movements are occurring among Palestinians and other subjugated groups, color, class, gender, ethnicity, ability, and religious and cultural divides are increasing (Meshulam & Apple, 2010).

Immigrant issues may lessen where there are ethnic and religious affinities, as in Pakistan, Israel, and Saudi Arabia. In Pakistan, the major immigrants are from Afghanistan, Burma, Iraq, Iran, and Somalia (Issa, Desmond, & Ross-Sheriff, 2010). Pakistan hosted Afghan refugees and welcomed them because they shared the same Islamic traditions. However, illegal immigrants sometimes face harassment, blackmail, and deportation from local governments.

Asia. East Asian countries, such as South Korea and Taiwan, have become multiethnic countries following their recent (since the 1990s) rapid industrialization. In addition, the visibility of foreign immigrants increased due perhaps in part to Korea's and Taiwan's low birth rate and citizens' avoidance of difficult, dirty, and dangerous jobs. The departure of young women from rural farm life to urban areas also provides a need for immigrants (Kang, 2010). As a result of industrialization and rural urban migration, many foreign brides are brought to South Korea and Taiwan to marry low SES (socioeconomic status) men in rural areas, work in the farms, give birth to Korean/Taiwanese babies, and care for the elderly (Bellenger et al., 2010; Choi, 2010; Kang, 2010).

Recent immigrants to China, Vietnam, Indonesia, Thailand, and the Philippines suffer from assimilation difficulties, patriarchy, and nationalism, although these countries state that they support multiculturalism and desire to eliminate differences. The educational systems in these countries emphasize concerns rooted in ideas of racial purity (Kang, 2010). Similar trends toward racial homogeneity and ethnocentric superiority also exist in Japan. Japan was well known as an imperialistic and colonialist power before World War II. During the period of Japan's colonial invasion in Asia, many Korean, Chinese, and Taiwanese laborers were brought to Japan and remained there. When Japan became one of most powerful world economies after World War II, many ethnic groups, including Chinese, Koreans, Taiwanese, Malays, Indians, Filipinos, and ethnic groups of Eastern European origin, immigrated to its islands. Until well into the 1980s and 1990s, immigrants were not granted citizenship or social rights in Japan (Kashiwazaki & Akaha, 2006). Now the Japanese government is grudgingly acknowledging the global trend toward multiethnic and cultural immigration.

Unlike Japan, Korea, and Taiwan, China has long been a multiethnic society trading with African, North American, Asian, and European countries. China has been a country of immigrants since it opened its doors to the West in the mid-eighteenth century (Chung, Qi, & Hou, 2010). Due to its economic growth and political stability in the past few decades, China has seen sudden increases in foreign investors, foreign students, and illegal immigrants from North Korea. China has begun to take in the skilled engineers and others with the expertise that will allow it to modernize and integrate with the world economy. Recent laws enacted show an attempt to control and monitor the influx of foreigners in order to prevent competition within local job markets. Yet, China needs to open up and bring in more foreigners "to enhance economic reform and further integration into the world economy" (Chung, Qi, & Hou, 2010, p. 357).

Overall, the general trend is that when the host country sees immigrants with highly developed social capital and trained skills, these immigrants are allowed to lead decent lives (Elliott et al., 2010). These skilled immigrants are seen as a resource that will help advance the host country. Conversely, illegal immigrants and unskilled foreigners are perceived as a drain on economic resources; thus their treatment has become less than humane in terms of working conditions, educational options, welfare benefits, and their general social and emotional well-being. Some undocumented/illegal immigrants are met with contempt and constant threat of deportation. Yet in all cases of immigrant populations, immigrants are subject to being paid less than native citizens, become terrorist suspects if they are from the Middle East or have a Muslim background, and are perceived as depriving mainstream citizens of competitive jobs in scarce job markets. Immigrants, in general, are facing harassment, exploitation, and racism.

Multicultural Education: A Solution to Global Injustices

Migration is a common global phenomenon. There is a constant flux of people moving all over the world as a result of globalization (Kang, 2010). Immigrants have come to and stayed in the host countries because they have economic opportunities and options that are greater there than in their homelands (Segal et al., 2010). However, the overall picture across the globe demonstrates that immigrants in destination countries are not treated with the same benefits and social rights as are native citizens. Labeled as undesirable people in the destination countries, refugees and unskilled and undocumented immigrants suffer exploitation and the threat of deportation in many parts of the world. The general tendency is that many immigrants (except in certain "desired" classes of skilled workers) are not only facing downward occupational mobility but are also made to feel the social and emotional humiliation of being second- or third-rate citizens in the host countries. Economic globalization may have led to "wider inequality, deeper division, and a dangerous era of distrust and rising tension" between immigrants and host country citizens (Green & Griffith, 2002, p. 65).

We argue that this trend of harassment, abuse, and threats to human rights in vulnerable immigrant populations is antithetical to the principles of global and multicultural education. Upholding xenophobia and ethnocentricity and idealizing racial purity and superiority in the global age are gross injustices for people who simply want to improve their economic options in their new country, even when "the less skilled immigrants are necessary for the functioning of the economy!" (Elliott et al., 2010, p. 460) and take the menial jobs that the native citizens typically shun.

As mentioned earlier, when schools function to serve and preserve the dominant groups' cultural values and knowledge, schools and teachers must take the initiative to promote cultural pluralism and multicultural education (Apple, 2000; Banks, 1993; Banks, 2002; Nganga & Kambutu, 2009). With increasing global communication, trade and businesses, facility of transportation, and the instant information exchange via technology, "the world is moving toward a global culture, and perhaps a global race" (Segal et al., 2010, p. 460). The longer the old thinking and traditional deficit views remain without broader and global perspectives, the more the current practices of inequality and injustice on fundamental human rights are exacerbated. In this light, schools and teachers should play a pivotal role in involving multiple perspectives and global and multicultural education at all levels and settings.

According to Brown and Kysilka (2009), living effectively with multiple views and foreign-born immigrants and competing for economic resources have been the most important factors for immigrants in the current global diaspora. Multicultural and global education emphasizes a process in which individuals develop knowledge, skills, attitudes, and behaviors to participate in a multicultural society on both the local and national levels (Banks, 2002; Blaut, 1993; Merryfield, 2002; Nganga & Kambutu, 2009). Global education highlights the importance of cosmopolitan citizenship, emphasizing world problems and the interconnectedness of the global economy. In addition, when global education is combined with multiculturalism, it embraces social justice and human rights on a worldwide scale, as Brown and Kysilka (2009) state:

> Multicultural and global education can be seen as the educational process of acquiring certain knowledge, skills, and values to participate actively in a complex, pluralistic, and interconnected world society and to work together for change in individuals and institutions in order to make that world society more just and humane. (p. 11)

To help immigrants cope with the societal burdens of neocolonialism and neoliberal capitalism that benefit those who occupy already privileged social positions, a global justice movement should start with a global and multicultural curriculum, schools, and teachers. The global and multicultural education should consist of five components (Brown & Kysilka, 2009): 1) understanding of social living in group; 2) understanding of the Other; 3) understanding of interrelatedness and interdependence; 4) development of skills in living with diversity; and 5) adjustment to changes for the future. To understand Others, live harmoniously with Others, and promote human connections, schools and teachers must adopt curriculum content that includes plural cultural views and materials (e.g., multicultural children's literature, world history and geography, worldwide sports, music, media texts, and entertainment) rather than using dominant monocultural texts and activities exclusively. Community-building and relationship-forming activities would also help students to accept other peoples' identities and encourage cultivating friendship among ethnically, racially, and culturally different "Others" (Choi, 2010). Hiring more administrators and teachers of different linguistic, cultural, and racial backgrounds can also benefit all students because it has been shown in extensive research (e.g., Banks, 1993; Lee, 1993; Antonio; 2001; Jayakumar, 2008) that people of color and different backgrounds bring with them diversity and act as role models to all students. Currently, educational professionals are largely from European and American backgrounds, and the curricular content reflects the selective viewpoint of the dominant culture in the United States and other economically powerful nations (Apple, 2010; Choi, 2010). Thus, students receive only a monolingual and monocultural perspective. To recognize injustice is good, but we (educators) should take a proactive stance and learn more about other cultures to improve the well-being of all. Our goal in the global age is to educate all individuals who are able to function effectively in ethnic cultures nationally and internationally (Banks, 2002).

Conclusion

In this chapter, we examined the meaning of globalization, global economies, and immigration. That people migrate to have more job opportunities for a better life is a natural process involving multiple and complex dimensions—economic, political, social, and human. We looked at the challenges posed to other people caused by globalization. From this, we concluded that global and multicultural education can be used to resolve global injustices through both a diversification of curriculum and a recognition of global interdependence. In reviewing global and multicultural educational goals, we recognized the need to uphold human rights and equity. Education can go only so far if policies are set and practices are based on deficit views of culturally and racially different peoples. At this point, we do need to see policy makers develop a comprehensive understanding of immigrant needs and to amend policies that see immigrants as a drain on their economies instead of as a resource. According to the International Labour Migration ("Illegal Acquisitions," 2010) migration is an essential and inevitable component of the global economy. Every state should develop a system to manage immigration for the benefit of both immigrants and the states concerned.

We also discussed globalization in the context of global market and free-trade treaties and their effects on developing and underdeveloped nations. These social issues can be addressed through using a global and multicultural education to educate world citizens about the need to coexist globally. Global and multicultural education can play an active role in promoting social justice through engaging learners at all levels and challenging issues of power and control (Brown & Kysilka, 2009). Because discrepancies exist in the power and control among the haves and have-nots; we need to bridge that gap. We educators must approach immigrant students with a better understanding of their backgrounds (as well as our own), and adjust our knowledge base, skills, and attitudes by working with and bringing in both dominant and subjugated views.

References

Akokpari, J. (2006). Globalization, migration, and the challenges of development in Africa. *Perspectives on Global Development and Technology*, 5(3), 125–153.

Anan, K. (2004). *Address to European Parliament upon receipt of Andrei Sakharov Prize for Freedom of Thought, Brussels*. Retrieved from http://ww.un.org/apps/sg/sgstats.asp?

Anderson, S. (2010). *Immigration: Greenwood guide to business and economics*. Santa Barbara, CA: Greenwood.

Antonio, A. L. (2001). The role of interracial interaction in the development of leadership skills and cultural knowledge and understanding. *Research in Higher Education*, (42), 593-617.

Apple, M. W. (2000). *Official knowledge: Democratic education in a conservative age* (2nd ed.). New York: Routledge.

Apple, M. W. (2004). *Ideology and curriculum*. New York: RoutledgeFalmer.

Apple, M. W. (Ed.). (2010). *Global crises, social justice, and education*. New York: Routledge.

Bacon, D. (2008). Displaced people: NAFTA's most important product. *NACLA Report on the Americas*, 41(5), 23-27.

Banks, J. (1993). Multicultural education for young children: Racial and ethnic attitudes and their modification. In B. Spodek (Ed.), *Handbook of research on the education of young children*. New York: Macmillan.

Banks, J. A. (2002). *An introduction to multicultural education* (3rd ed.). Boston: Allyn & Bacon.

Bauer, T., & Zimmermann, K. (1998). Causes of international migration: A survey. In P. Gorter, P. Nijkamp, & J. Poot (Eds.), *Crossing borders: Regional and urban perspectives on international migration* (pp. 95-127). Aldershot, England: Ashgate.

Bellenger, D., Lee, K. H., & Wang, H. Z. (2010). Ethnic diversity and statistics in East Asia: Foreign brides surveys in Taiwan and South Korea. *Ethnic and Racial Studies*, 33(6), 1108-1130.

Bhuyan, R. (2010). Reconstructing citizenship in a global economy: How restricting immigrants from welfare undermines social rights for U.S. citizens. *Journal of Sociology & Social Welfare*, 37(2), 63-85.

Blaut, J. M. (1993). *The colonizer's model of the world*. New York: Guilford Press.

Brown, S. C., & Kysilka, M. (2009). *Multicultural and global education*. Boston: Pearson.

Choi, J. S. (2010). Educating citizens in a multicultural society: The case of South Korea. *The Social Studies*, 101, 174-178.

Chung, K. W., Qi, J., & Hou, W. (2010). China: A new pole for immigration. In U. A. Segal, D. Elliott, & N. S. Mayadas (Eds.), *Immigration worldwide: Policies, practices, and trends* (pp. 352-362). Oxford, England: Oxford University Press.

Don, R., & Gretchen, H. (1991). *Food self-reliance in poor countries: GATT should support it, too.* Retrieved from http://web.ebscohost.com.proxy.uwlib.uwyo.edu/ehost.

Elliott, D., Segal, U. A., & Mayadas, N. S. (2010). Immigration worldwide: Themes and issues. In U. A. Segal, D. Elliott, & N. S. Mayadas (Eds.), *Immigration worldwide: Policies, practices, and trends* (pp. 451-464). Oxford, England: Oxford University Press.

Gibson, M.A., & Rojas, A. R. (2006). Globalization, immigration, and the education of "new" immigrants in the 21st century. *Current Issues in Comparative Education*, 9(1), 69-76.

Green, D., & Griffith, M. (2002). Globalization and its discontents. *International Affairs*, 78(1), 49-68.

Haas, H. D. (2010). The internal dynamics of migration process: A theoretical inquiry. *Journal of Ethics and Migration Studies*, 36(10), 587-617.

Hettne, B., Inotai, A., & Sunkel, O. (1999). *Globalism and the new regionalism*. London: Macmillan.

Illegal acquisition of citizenship behind crackdown on immigrants in Kenya. (2010, February). *BBC Monitoring International report*. Retrieved from International Labour Migration (2010). International labour migration: A rights-based approach. International Labour Office, Geneva, Switzerland.

Issa, S.S., Desmond, G., & Ross-Sheriff, F. (2010). Pakistan: Refugee history and policies of Pakistan: An Afghan case study. In U. A. Segal, D. Elliott, & N. S. Mayadas (Eds.), *Immi-*

gration worldwide: Policies, practices, and trends (pp. 171-188). Oxford, UK: Oxford University Press.

Jayakumar, U.M. (2008). *Can higher education meet the needs of an increasingly diverse and global Society? Campus diversity and cross-cultural workforce competencies.* Retrieved from http://www.edreview.org/harvard08/2008/wi08/w08jayak.htm

Kambutu, J., & Nganga, L. (2008). In these uncertain times: Educators build cultural awareness through planned international experiences. *Teaching and Teacher Education,* 24(4), 939-951.

Kang, S.W. (2010). Multicultural education and the rights to education of migrant children in South Korea. *Educational Review,* 62 (3), 287-300.

Kashiwazaki, C., & Akaha, T. (2006). *Japanese immigration policy: Responding to conflicting pressures.* Retrieved from http://www.migrationformation.org/Feature/display.cfm?ID=487

Korovkin, T. & Sanmiguel-Valderrama, O. (2007). Labour standards, global markets and non-state initiatives: Colombia's and Ecuador's lower industries in comparative perspective. *Third World Quarterly,* 28(1), 117-135.

Leighton Schwartz, M, & Notini, J. (1994). *Desertification and migration: Mexico and the United States.* San Francisco: Natural Heritage Institute.

Lee, O. (1993). Equity for linguistically and culturally diverse students in science education: A research agenda. *Teachers College Record,* 105, (3), 465-489.

Lingard, B. (2007). Deparochializing the study of education: Globalization and the research imagination. In K. Gulson & C. Symes (Eds.), *Spatial theories of education: Policy and geography matters* (pp. 233-250). New York: Routledge.

Litz, D. (2011). Globalization and the changing face of education leadership: Current trends and emerging dilemmas. *International Education Studies,* 4 (3), 47-61.

Mabogunje, A.L. 1970. Systems approach to a theory of rural-urban migration. *Geographical analysis,* 1 (2), 1-18.

Mander, J., Baker, D., & Korten, D. (2001). Does globalization help the poor? *IFG Bulletin,* 1(3). Retrieved from http://www.thirdworldtraveler.com/Globalization/DoesGlobalizHelpPoor.html

Massey, D. S., Arango, J., Hugo, G., Kouaouci, A., Pellegrino, A., & Taylor, J. E. (1998). *Worlds in motion: Understanding international migration at the end of the millennium.* Oxford, England: Clarendon Press.

Merryfield, M. M. (2002). The difference a global educator can make. *Education Leadership,* 60(2), 18-21.

Meshulam, A., & Apple, M. W. (2010). Israel/Palestine, unequal power, and movement for democratic education. In M. W. Apple (Ed.), *Global crises, social justice, and education* (pp. 113-162). New York: Routledge.

Michalowski, I. (2010). Immigration to France: The challenge of immigrant integration. In U. A. Segal, D. Elliott, & N. S. Mayadas (Eds.), *Immigration worldwide: Policies, practices, and trends* (pp. 79-94). Oxford, England: Oxford University Press.

Nganga, L., & Kambutu, J. (2009). Teaching for democracy and social justice in rural settings: Challenges and pedagogical opportunities. In A. L. Groenke & J. A. Hatch (Eds.), *Critical pedagogy and teacher education in the neoliberal era* (pp. 191-204). Montreal, Quebec, Canada: Springer.

Rajiman, R., & Kemp, A. (2010). Israel: The new immigration to Israel: Becoming a de facto immigration state in the 1990s. In U. A. Segal, D. Elliott, & N. S. Mayadas (Eds.), *Immigration worldwide: Policies, practices, and trends* (pp. 227-243). Oxford, England: Oxford University Press.

Segal, U. A., Elliott, D. & Mayadas, N. S. (2010), *Immigration worldwide: Policies, practices, and trends*. Oxford, England: Oxford University Press.

Somerville, W., & Cooper, W. (2010). Immigration to the United Kingdom. In U. A. Segal, D. Elliott, & N. S. Mayadas (Eds.), *Immigration worldwide: Policies, practices, and trends* (pp. 124-134). Oxford, England: Oxford University Press.

Trimikliniotis, N., Gordon, S., & Zondo, B. (2008). Globalization and migrant labor in a "Rainbow Nation": A fortress South Africa? *Third World Quarterly*, 29(7), 1323-1339.

United Nations Population Division (UNPD) (2006). *Trends in total migrant stock: The 2005 revision*. New York: Department for Social and Economic Affairs, UNDP.

Wickramasekara, P. (2008). Globalization, international labour migration and rights of migrant workers. *Third World Quarterly*, 29(7), 1247-1264.

Zohry, A. (2010). Egypt: Immigration in Egypt. In U. A. Segal, D. Elliott, & N. S. Mayadas (Eds.), *Immigration worldwide: Policies, practices, and trends* (pp. 323-334). Oxford, England: Oxford University Press.

Zufiaurre, B. (2006). Social inclusion and multicultural perspectives in Spain: Three case studies in northern Spain. *Race, Ethnicity and Education*, 9(4), 409-424.

CHAPTER FIVE

Preparing Teachers for *Global* Citizenship: Perspectives from One Caribbean Island

Karen Thomas-Brown

Historically the islands of the Caribbean are highly globalized, yet localized places, due in part to multiple phases of globalization. Today, a second wave of globalization has affected the political, economic and social structures of many of these islands through neoliberalist policies and post-colonialism, among a plethora of other theories, policies and approaches. Within this context, the education system in the Caribbean in general and in Jamaica in particular was founded on transformational pedagogies (Freire, 2000) and has historically adapted and readapted to societal needs as the political and economic fortunes of the islands wax and wane. This chapter examines the impacts of and response to a relatively new migration phenomenon in the Caribbean islands such as Jamaica: the acceleration of migrations of teachers in response to recruitment drives that have originated in the United States, Canada and the United Kingdom starting in the mid-1990s. These recruitment drives have attracted the most qualified, highly skilled and experienced teachers and nurses in exchange for the possibility of improved economic prospects in developed countries. The chapter shows that the islands' responses to this teacher emigration range from lack of awareness, improved local working conditions, changed national curricula, increased wages and salaries for teachers and facilitation of the emigration process (Degazon-Johnson, 2007, 2008; Ministry of Education of Jamaica, 2004). This chapter takes a multitheoretical approach to the notion of global education as it elucidates the issues of the impact of globalization on teacher training and the education system of a group of small developing states. The chapter also expounds on specific examples of critical, crisis and transformational pedagogies (Freire, 2000) as those who work in teacher-training institutions encourage their students and graduates to be global citizens. Finally, the chapter shows that the incorporation of social studies principles into all aspects of teacher training on the islands is in pursuit of creating the global citizen/teacher.

Introduction

The need to prepare future teachers who are globally aware is as significant a need in the Caribbean as it is in the United States and in other parts of the world. The islands of the Caribbean are highly globalized spheres, having experienced their first incorporation into the global economy through the Columbian voyages of the 1400s. These islands experienced further global incorporation through colonial imperialism and later through neoliberalism. While all these phases of global incorporation have focused primarily on the economic and political fortunes of many of the islands, the inherent policies have had significant effects on the mandates, structure, function and purpose of education. Historically, education on many of the Caribbean islands was instrumental in meeting the needs of the metrople; however, with the post-1950s independence of many islands such as Jamaica, the focus of education shifted to meet the needs of the newly independent countries. Since the mid-1970s, islands such as Jamaica experienced the rise of the neoliberal age, and with this rise, the need to prepare more global citizens, given new, more globally competitive economic realities and more capitalistic political/governmental expectations. Hence, the general notion that education mirrors society, and in a rapidly globalizing world, it is imperative that education begins to meet the needs of this globalized society. This is especially pertinent in the context of many in these small island states. Ground zero for such discussions, debates and discourses has to be teacher preparation programs, given the substantial and yet grassroots role of teachers in shaping future citizens.

In recent years islands such as Jamaica have experienced increased emigration of their qualified teachers. This emigration has raised several important questions pertaining to the curriculum mandates on the island; the preparation levels of the migrating teachers; brain drain, brain circulation or remittances and return flow gains; and the impact of the losses in teachers on the quality of education on the island. These are the concerns that this chapter attempts to address within the context of existing discourses in global education and teacher preparation. Therefore, how do we prepare teachers to be capable of instilling in our students ideas that allow for the growth of productive, informed, global citizens who are able to independently deliberate on existing and emerging global issues and topics? How do we equip global citizens with the skills to take action where necessary, while still contributing to the greater good of their island nation and the globe?

The Importance of Preparing Teachers Worldwide for Global Education (Global Issues and Trends)

Global citizenship is an analytical lens to construct conversations about teacher preparation programs globally. The arguments supporting the preparation of teachers everywhere for global education may be subsumed within the notion of global citizenship and the need to be familiar with one's national as well as global contexts, connectivity and implications. According to Bremer (2007, p. 41) "international education matters and philanthropic organizations are taking notice by funding projects that support the mission and goals of international educators." "Global educators help students explore other cultures through literature, history, news, and Web sites for other parts of the world" (Merryfield, 2002, p. 19). Globalization's impact on higher education raises several fundamental questions; chief among these is how to serve the common good within the context of global norms (Segrera, 2010). Learning and work can no longer occur within the confines of national contexts; today's global marketplace demands workers who are internationally competent and equipped with new and emerging skill sets in the pursuit of solving global problems. Gaudelli (2009) quoting May confirmed this point by noting that capitalist globalization fosters solidarity while offering workers internationalism as the path to their futures. This, according to Gallanvan (2008), is rooted in the essence of global education, the content and processes of which involve cultivating a global perspective, subsumed within which are the attributes for ethical, pluralistic and interdependent deliberations needed to efficiently navigate the world and be active citizens.

Within the Caribbean as well as in many other areas, the primary intersection of globalization and education begins with three main measures of globalization: trade; private cross-border financial flows as well as foreign direct investments (FDI); and migration. These measures influence the competitiveness and adaptability of a place to globalization through years of schooling (elementary/primary and high school/secondary school), and through vocational, college-level and international education. Trade and FDI have a positive effect on education and training. This is both direct, through more training, and indirect, through the available resources. However, the overall effects are more positive in countries that are already relatively well endowed with educational facilities to start with (te Velde, 2008). The complexity of international migration results in both positive and negative outcomes. The positive outcomes may be seen in the network effects on trade in goods and services and remittances, whereas emigration from smaller countries results in apparent capacity losses in specific professions such as teaching. For many developing countries, vocational education and college-level education are useful in attracting FDI; however, according to te Velde (2008) they need to be appropriate and include engineering and other technical skills. Examples from Thailand and South Korea demonstrate that a pool of well-educated nationals abroad can act as a source for exported goods and services and as a source for diaspora investment back into the home country, much of which goes towards daily living expenses and education. The same is true in Jamaica. Gaudelli (2009) purports that "global citizenship curriculum lacks natural consistency" (p. 77), making the need to examine notions of teaching for globalization much more pertinent. The core dimensions of global education, according to Zong et al. (2008) include attention to such global issues as justice, equity and sustainability. The spatial dimension focuses on global-local flows and perspectives; the temporal dimension focuses on the interconnected nature of places over time; and the process dimension is the one most critical to teacher educators since it focuses on participatory actions, pedagogy and the actualization of global citizenship.

What then is the purpose of global education and where should the primary focus lie? Should there be an emphasis on trade, growth, poverty reduction, nation-building, sustainable development, national and human security, terrorism or HIV/AIDS? Should education be viewed as a commodity, and be traded in services? This is often the case in several Caribbean islands, where many college-educated individuals are encouraged to migrate with the assumption that they will still contribute to the islands' economies through remittance. Within this context, what should the role of teacher education be and what do teachers need to know in order to aid students as they navigate the world? According to Holden and Hicks (2007), longstanding global issues include poverty, environment, conflict and social justice and these affect the world within which students live (see also Hicks & Holden, 1995; Hutchinson, 1996; Oscarsson, 1996). Merryfield (2002) notes that "by examining different points of view . . . students develop the habit of looking for and considering other perspectives, especially those of people of minority cultures or from other continents" (p. 19).

Regardless of the path a student takes in the world, be it vocational training or college- and university-level training, he or she requires the communication and leadership skills and knowledge base to function in a globalized world. According to te Velde (2008, p. 6), "the quality of education and training determine whether and how countries can participate in the process of globalization, such as global value chain, fragmentation, increased trade in final products, and migration." According to Tye in Holden and Hicks (2007, p. 14) "global education involves learning about those problems and issues which cut across national boundaries and about the interconnectedness of systems, cultural, ecological, economic, political, and technological." Therefore, reflected within global education is a demonstration of understanding of and empathy and appreciation for those perceived to be "others" and "neighbors." Such categorizing allows for more insightful interactions that penetrate cultural, geographic, economic and social obstructions.

Studies show that only a small number of teacher preparation courses globally attempt to promote global education (Holden & Hicks, 2007). The nature of teacher training in many developed countries allows for the infusion of global education and global perspectives by instructors and professors. The focus themes tend to be reformist education and wealth, power and rights disparities. This, according to Segrera (2010), allows for a broad social mandate that provides more relevant and innovative curricula. This corresponds with the views of Alazzi (2011), Gaudelli (2009) and Merryfield (2002), who noted that significant elements of global education include knowledge and inquiry into global interconnectedness and issues, and open-mindedness that facilitates sensitivity to varied perspectives. This in essence allows for the recognition of biases, stereotyping and exoticism and is best practiced within the context of intercultural experiences, understandings and competences.

Globally, there is a trend towards an increased push for global education given debates and increased attention paid to issues of citizenship, race equality, sustainability and climate change, political conflicts, global economic flux and the social and cultural changes brought about by increased diversity. This brings to the fore contestations and problematizing about the impact of these changes on notions of identity and sense of belonging and where these conversations should be facilitated. This, according to Gaudelli (2009, p. 75), is a "shift evident in the past 60 years away from national citizenship and towards standing as a person irrespective of national affiliations or the lack thereof." Globally, teacher education takes place in universities, schools of education or specialized teacher training colleges. On average, a teaching degree takes four years to complete or a period of graduate training if one's bachelor's degree is not in teaching. Of significance is the fact that these programs focus on national and state curricula, planning, assessment and classroom behavior management. The infusion of global education themes is dependent on the mandates and mission of the university or college and is strongly influenced by regulatory boards such as the department of education in each state, the teacher training agency (TTA) in England and the joint board of teacher education in some Caribbean islands. These bodies decide the standards against which all future teachers must be measured in order to gain teacher certification. The general trend also indicates that those individuals seeking teacher certification major and or minor in one or more subjects. This in effect limits the opportunities for them to be exposed to the many themes of global education and the importance of global citizenship. This is within the context of the post-independence period in which several Caribbean islands have diligently changed school curriculum requirements and high school/secondary-level standardized examinations to reflect Caribbean and island-specific themes, topics and issues. This is essentially a post-colonial shift towards a more nation-alist focus that was seen by some governments as more relevant. This, according to Gaudelli (2009), represents the centrifugal force that pushes back against some aspects of globalization in favor of regional character. This, for many post-independent Caribbean islands is an argument for greater self-determination.

The Holden and Hicks (2007) study found that most future teachers wish to learn more about global issues because they think this would allow them to better facilitate their own students when such discussions, questions and topics are raised because there is a fundamental link between active citizenship and global awareness. Similar studies from Banks (2008) and Hanvey (1976) found that many future teachers, while enthused about infusing the global perspective in their future teaching, were hesitant about how to and when to do so, particularly as it pertains to sensitive topics that have inherent biases. This concern expressed by future teachers is a direct outcome of the limited and or fragmented infusion of the global perspective into their teacher preparation courses. While in the United States students seem to have opportunities to be exposed to global perspectives and themes in the courses they take, there are regulatory groups and time constraints that influence how much of this information is made available to these future teachers in usable form for classroom practice. This is important because it calls into question notions of academic mobility within the context of continued national structure.

The emerging issues in global education pertain to some modification of the roles and pedagogy of teachers and to increased exposure to and training in the use of new and emerging information technologies. Other issues relate to the certification process and how best to organize these so that national and state interests and needs are maintained, but the infusion of global issues allows the teacher to be mobile academically. In this sense, academic mobility allows for one's credentials earned in one country or state to be transferable to other academic and country or state settings without the requirement of complete retraining. The challenge that faces many educators is the absence of an agreed-upon definition of global competence. How do we know our teachers and the students they serve meet basic standards of global preparedness? According to the American Council for Education (ACE) report (1988) in Hunter (2004, p. 8), "To be globally competent, one must have four or more international college courses and have an unspecified ability to speak a foreign language." This is a limited view of such an important concept because it does not meaningfully provide for the complexity of global education, experience and competencies aforementioned. Global competence is critical not only to the success of cross-border workers; global competence is the possession of cultural capital that also allows one to understand, empathize with and communicate in an array of cultural settings. It is the knowledge and awareness of global topics, trends, events, issues and debates that allow one to be able to effectively communicate across geographic, cultural and linguistic borders.

This for Merryfield (2002) and Alazzi (2011) allows these individuals to be open minded; resistant to stereotyping; and able to communicate across cultures while establishing connections between themselves and other countries, international organizations and cultures. Thus this allows students to participate more meaningfully as global citizens.

The Context of Social Studies Classrooms

Bromley (2009) wondered about the extent to which the principles of global education have become more infused in social studies, civics and social science education since World War II. To find out, Bromley evaluated high school civics, history and social studies textbooks published since 1970. The outcome of this study indicates that "there is a longitudinal trend towards greater emphasis on universalism and diversity across a broad range of countries" (p. 37). There is increased visibility of global citizenship in school curricula globally, thus a cosmopolitan emphasis in civics education textbooks. There is also an increased mentioning of global citizenship in several (approximately twelve) state curricula. However, knowledge of the extent to which this content translates into classroom practice is still lacking.

Policies That Affect Multiple Forms of Citizenship

The nature of education in Jamaica and its connection to the global may be contextualized using Gaudelli's (2009, p. 70) Heuristics of Global Citizenship: in Figure 1, the X-axis is the tangible-imaginary and represents the extent to which global citizenship is suffused into national institutions and processes. The Y-axis represents competitive-cooperation and indicates the level of competitiveness or cooperativeness of citizenship actions. In Jamaica, the vision of global citizenship intersects to produce competitive-tangible-cooperative individuals, given the nature of K–12 curricula and the outcome. Despite this, the desired outcome is a global citizen who is more cosmopolitan. Citizenship is a contested construct socially, politically and academically, hence the implications of responsibilities to and rights of the state. Knight Abowitz and Harnish (2006) and Urry (1999) in Rapoport (2009) identified discourses and models of citizenship. These include civic republican; liberal; feminist; reconstructionist; cultural; queer; transnational discourses and cultural; minority; ecological; cosmopolitan; consumer; and mobility categorizations. For Westheimer and Kahne (2004a) three types of citizenship are important: the personally responsible citizen, the participatory citizen and the justice-oriented citizen. According to these writers it is possible to pursue all three visions of citizenship; however, there will be conflicts because the emphasis placed on individual character and behavior as an example may conflict with the need for collective and public sector efforts.

Figure 1. Jamaica Added to the Gaudelli's Heuristics of Global Citizenship

```
                        Competitive              Curriculum
     Actual Curriculum
                                                   Vision of Citizen
              National /Regional      Marxist
                                                 Vision of Civic
Vision of Civic

           Neoliberal
Tangible                                               Imaginary




            Vision of Citizen

                                    Desired Curriculum
     World Justice and Governance       Cosmopolitan
                        Cooperative
```

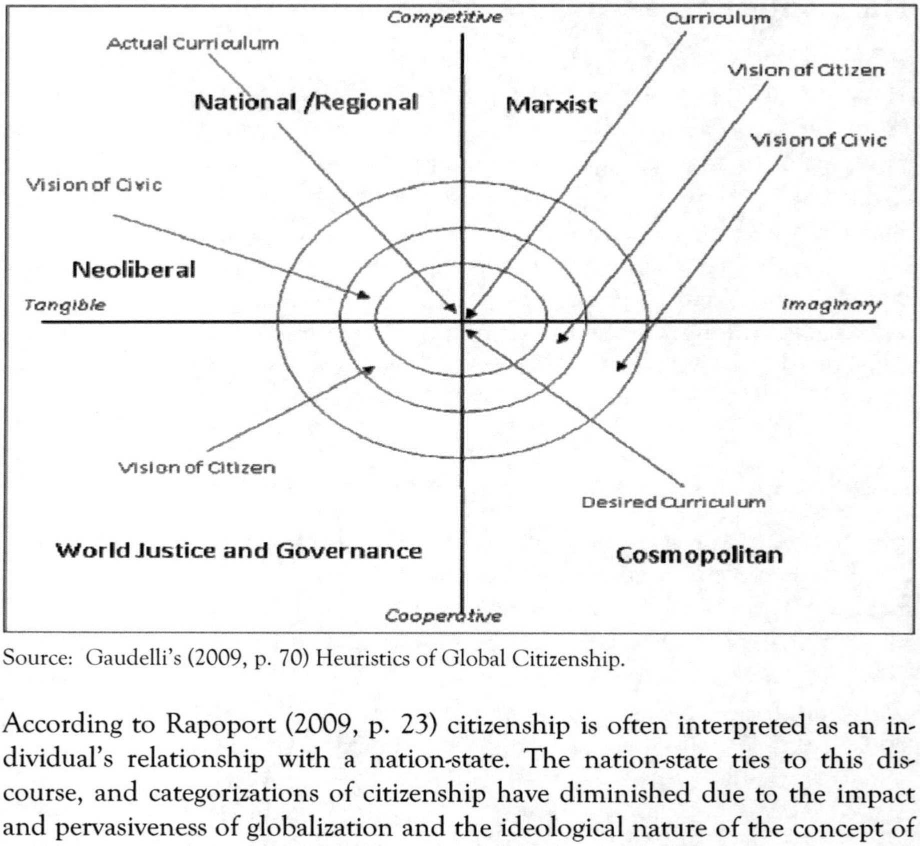

Source: Gaudelli's (2009, p. 70) Heuristics of Global Citizenship.

According to Rapoport (2009, p. 23) citizenship is often interpreted as an individual's relationship with a nation-state. The nation-state ties to this discourse, and categorizations of citizenship have diminished due to the impact and pervasiveness of globalization and the ideological nature of the concept of the nation-state and its ancillary nation-building. Hence, the concept of the nation-state is a vulnerable one, because of a state's implicit predisposition to change in response to internal and external forces. In the United States, the research literature on schools and citizenship proves otherwise. Knight Abowitz and Harnish (2006) as well as Westheimer and Kahne (2004a) indicate that traditional/personally responsible views of citizenship are dominant discourses in schools. Furthermore, the Westheimer and Kahne (2004b) study found that citizenship education is rarely discussed in the public schools (in this case, in Chicago), and fear of the lack of citizenship education is typical. The transformative nature of the nation-states and notions of citizenship is precipitated by the rise in global consciousness and global citizenship. Tully (2008) and Rapoport (2009) added to this debate by noting that cosmopolitan citizenship is at minimum similar to global democratic citizenship. The cosmopolitan approach is multidimensional, and it emphasizes the "interconnected society and culture that is unbounded by the political territory of nation-state" (Bromley, 2009, p. 34). The rights of all human beings, as a global citizenship theme, figure prominently in the work of Pogge (2008), Levy and Sznaider (2004) and Beetham (1999). In addition, a shift has taken place from a nationalistic focus to trans-border changes and their impacts on individual choice, tolerance, vulnerability and social participation, which are also considered important. Cosmopolitan education "involves a willingness to tolerate, celebrate, engage openly with, and even seek out diverse social and cultural experiences" (Bromley, 2009, p. 34).

Cosmopolitan or global citizenship is a critical component of citizenship education. Given the transformative nature of globalization, cosmopolitan and global education provides students with exposure to multiple and divergent perspectives. Given the rise in transnational migration, there is increased racial, cultural, ethnic and religious diversity. This increased diversity in many places precipitates the need to teach about the "other." Critical within this is teaching about and to the needs of minority groups as well as moving beyond "American exceptionalism." Kammen (1993, p. 6) defines American exceptionalism as "the notion that the United States has had a unique destiny and history, with distinctive features or an unusual trajectory"; therefore, it is fundamental to society's sense of its own identity. However, VanSledright (2008, p. 110) added to the discourse by indicating that "despite America's status as a nation of immigrants (both voluntary and involuntary) and persistent social struggles to negotiate vast cultural differences among its citizens, what gets transmitted to students in social studies classrooms is often an uncomplicated narrative of national development and progress." This unifying narrative of U.S. history is criticized by Kim (2004), who suggests that there is a failure to address the importance of multiple perspectives and interpretations in historical analysis. Inherent within this is the notion of Manifest Destiny, which, according to Kim (2004), is the historical embodiment of American exceptionalism. However, when social studies curricula fail to contextualize the origins and implications of Manifest Destiny, teachers and students are inclined to treat it as a self-evident universal construct. These are important concepts within teacher education, because they imply that teachers need to be aware of their status as global cosmopolitan citizens and be equipped with the relevant skill set to operate effectively and to inform their students.

Despite the recognized need, studies conducted by Lee and Leung (2006) in Hong Kong and Schweisfurth (2006) in Ontario, Canada, found that teachers face many difficulties in teaching about global and cosmopolitan citizenship because of locally focused curricula that are examination oriented and standardized and because of lack of training and limited administrative support. Consequently, what is often seen are debates about global and cosmopolitan citizenship at the college and university levels, with inadequate seepage of this discourse to teachers who operate at the practitioner level where guidance is needed to teach from a global perspective. According to Rapoport (2008) in a study conducted among teachers in Indiana, global citizenship is rarely mentioned in classrooms because most teachers assert that they cannot teach what they do not know. Other factors that negatively affect the inclusion of global principles in teaching include time-tabling of classes and the time constraints within which teachers must work. For Rapoport (2008, 2009), the fundamental challenges to teaching global citizenship are theoretical and terminological divergence within the context of nationalist and patriotic curricula.

Perspectives from One Caribbean Island: Jamaica

There have been accelerated emigrations of teachers from several Caribbean islands in response to recruitment drives that have originated in the United States, Canada and the U.K. starting in the mid-1990s. These recruitment drives have attracted the most qualified, highly skilled and experienced teachers in exchange for the possibility of improved economic prospects in developed countries. The islands' responses to this teacher emigration range from lack of awareness, improved local working conditions, changed national curricula, to increased wages and salaries for teachers and facilitation of the emigration process (Degazon-Johnson, 2007, 2008; Ministry of Education of Jamaica, 2004). It is important to note that emigration from the islands of the Caribbean is not a new phenomenon. Historically, the islands experienced extensive emigration during the pre- and early independence periods of the late 1950s to mid-1960s and during periods of political crisis (e.g., 1979–1981 in Jamaica). The negative impacts of emigration from the islands such as brain drain, loss of local earning power, negative multiplier effect, import dependency and human capital loss are well studied. The past decade has seen the Caribbean lose as much as 70 percent of its college-trained workforce to developed countries (Degazon-Johnson, 2007; Mishra, 2006). For example, between 2001 and 2003, Jamaica lost approximately one thousand teachers to the U.K.

alone. While acknowledging the human capital loss, several studies (Dodman, 2009; Timms, 2008) have pointed to the benefits that many of the Caribbean islands derive from the skilled migration of some of its educated population in the form of return flows (remittances, containers, other gifts and returning residents/circular migrants).

On the island, the general consensus among those who work in teacher-training institutions and in the university department devoted to teacher education is one of lack of inclusion in the recruitment and emigration of teachers. The advertisements that attract the best of the island's teachers are placed in local newspapers and on the Internet and no collaboration takes place between the local teaching institutions and the recruiting agencies. There is also a lack of knowledge among those who work in teacher training institutions about the recruitment protocols used as well as the extent to which the rights of the recruited teachers are considered during the recruiting process. Despite this, it is generally accepted that the migration experience enhances one's proficiencies and there are benefits for both the recruiting country and the home country of the teachers. Conversations with several college lecturers at Mico University College in Kingston, Jamaica, indicate some consensus about how the emigration of teachers is perceived. According to one lecturer, "We do not see globalization as a threat. We see it as a positive, widening scope that always enhances a person's development; the bicultural experience/exchange, whether temporary or permanent, can benefit countries." While such perspectives acknowledge the potential benefits of teacher emigration, there are no published studies to substantiate or refute these claims; also absent are studies that indicate the possible negative impact of teacher emigration on the quality of the island's K-12 education system.

According to Spring (2008), quoting the Global Commission on International Migration (2005), brain circulation is a process "in which migrants return to their own country on a regular or occasional basis, sharing the benefits of the skills and resources they have acquired while living abroad." This contextualizes some of the debates about teacher emigration in Jamaica. Common conversations indicate that many Jamaicans believe that with teacher emigration the island loses valuable human capital, while others focus on the return flows to the islands. Few discussions focus on the notion of brain circulation, which is relevant to those teachers who choose to return to the island after their contracts overseas end. These individuals convey an enhanced worldview that has the potential to positively affect their teaching. At this juncture, critical, crisis and transformational pedagogies (Freire, 2000; Austin, 1996) suggest that those who work in teacher training institutions encourage their students and graduates to be global citizens. This is not always the case, because the curricula taught are structured and assessed by one governing body (the Joint Board of Teacher Education) and the primary focus remains on nationalist themes to meet the needs of Caribbean-based standardized testing at the end of primary/elementary school and at the end of high school/secondary school. Therefore, it is the general consensus that global citizenship themes are not routinely infused in the curriculum, and when they are infused they are limited by the teacher's/lecturer's knowledge, perceptions and experience with globalization. Additional conversations with college lecturers indicate that there is a growing demand for local teachers to be more globally focused, more tolerant of those they perceive as different from themselves and able to communicate more clearly using a global language. According to Spring (2008), English may be considered the global language. Within the Jamaican context, while English is the language used in schools as well as being the island's official language, for many it is not normally their first language. Despite the fact that many Jamaicans' first interactions with English as a written language is in church or school, it is not taught as a second language, which presents some communication challenges for many future teachers. There is also consensus that given the extent of transnational migration and the large numbers of local teachers who emigrate to developed countries in response to recruitment drives, there should be some standardization of teaching certifications that would allow those who choose to emigrate the opportunity to transition more smoothly into the new country.

Similar to teacher educators in the United States, teacher educators in Jamaica are faced with the need to incorporate global citizenship principles into curricula and pedagogy as they respond to the transformational nature of globalization and increased global interdependence. Similar to what Zong (2009) thinks, there is the recognition that in an interconnected world, one's survival is directly tied to one's capacity to understand and effectively interact with changes that originate outside of national settings.

Implications for Teacher Education Programs

This topic is critical and timely for a many reasons, including the fact that as teacher training on the Caribbean islands adapts to the desires of their graduates to emigrate, several cultural conversations must be rearticulated to make these individuals more culturally adaptive and responsive to their potential work environments. This brings to the fore the pervasiveness of social studies education and the core of this subject, which is to encourage students to become "good citizens." Today, the incorporation of social studies principles into all aspects of teacher training on the islands is in the pursuit of creating the "global citizen/teacher": one who possesses the skill sets that make him or her an effective teacher in any country setting that he or she chooses to work in.

Globalization is reflected in two aspects of teacher education programs in many national systems: in the diversity of the student population that teacher candidates are required to support, and in the rhetoric of the proclaimed goals of these programs. The realization of the proclaimed goals, however, is influenced by historically and politically established local social and cultural factors. Program entry and exit requirements, description of major program components and attractiveness of teaching as a career in different societies are discussed, and the tensions between global and local aspects of teacher education illuminated. The education necessary to prepare college graduates to be globally competent lacks clarity, uniformity and direction. The future implication is to have these issues addressed in the curriculum and pedagogy of teacher education to the extent that practitioners begin to feel comfortable enough to attempt to infuse global and cosmopolitan citizenship into their daily teaching. This will require a rethinking of the nationalist and state focus of social studies, history and civics curriculum as well as exit standards for teachers.

References

Alazzi, K. (2011, December). Teachers' perceptions and conceptions of global education: A study of Jordanian secondary social studies teachers. *The Journal of Multiculturalism in Education, 7*, 1-19.

Beetham, D. (1999). *Democracy and human rights.* Malden, MA: Polity.

Bremer, D. (2007). Mission driven. *International Educator, 16*(3), 40-47.

Bromley, P. (2009). Cosmopolitanism in civic education: Exploring cross-national trends, 1970-2008. *Current Issues in Comparative Education, 12*(3), 33-44.

Degazon-Johnson, R. (2007, May). *Migration and Commonwealth small states—the case of teachers and nurses [Report].* Presented at the University of the West Indies Sir Arthur Lewis Institute of Social and Economic Studies (SALISES) and Commonwealth Secretariat conference, Barbados. Retrieved from http://www.thecommonwealth.org/files/182470/FilesName/UWISALISESLabourMarketConference_May2007_1.pdf

Degazon-Johnson, R. (2008). *Ethical recruitment standards: The case of the Commonwealth Teachers Recruitment Protocol.* Paper presented at the American Federation for Teachers Migration conference, Chicago. Retrieved from http://www.thecommonwealth.org/files/182468/FileName/AFTMigrationForumPresentation_July08_.pdf

Freire, P. (2000). *Pedagogy of the oppressed.* New York: Continuum International Publishing Group.

Gallanvan, N. (2008, November/December). Examining teacher candidates' view on teaching world citizenship. *The Social Studies,* 249-254.

Gaudelli, W. (2009). Heuristics of global citizenship discourses towards curriculum enhancement. *Journal of Curriculum Theorizing, 25*(1), 68-85.

Gaudelli, W., & Fernekes, W. (2004). Teaching about global human rights for global citizenship. *The Social Studies, 95*(1), 16-26.

Hicks, D., & Holden, C. (1995). *Visions of the future: Why we need to teach for tomorrow*. Stoke-on-Trent, England: Trentham Books.
Holden, C., & Hicks, D. (2007). Making global connections: The knowledge, understanding and motivation of trainee teachers. *Teaching and Teacher Education*, 23, 13–23.
Hunter, W. (2004). Got global competency? *International Educator*, 13(2), 6–12.
Hutchinson, F. (1996). *Educating beyond violent futures*. London: Routledge.
Kammen, M. (1993). The problem of American exceptionalism: A reconsideration. *American Quarterly*, 45(1), 1–43.
Kim, C. (2004). Imagining race and nation in multiculturalist America. *Ethnic and Racial Studies*, 27(6), 987–1005.
Knight Abowitz, K., & Harnish, J. J. (2006). Contemporary discourses of citizenship. *Review of Educational Research*, 76(4), 653–690.
Lee, W., & Leung, S. (2006). Global citizenship education in Hong Kong and Shanghai secondary schools: Ideals, realities and expectations. *Citizenship Teaching and Learning*, 2(2), 68–84.
Levy, D., & Sznaider, N. (2004). The institutionalization of cosmopolitan morality: The Holocaust and human rights. *Journal of Human Rights*, 3(2), 143–157.
Merryfield, M. (2002). The difference a global educator can make. *Educational Leadership*, 60(2), 18–21.
Ministry of Education of Jamaica. (2004). *Protocol for the recruitment of Commonwealth teachers [Report]*. Retrieved from http://www.thecommonwealth.org/shared_asp_files/GFSR.asp?NodeID=39311
Oscarsson, V. (1996). Young people's views of the future. In A. Osler, H.-F. Rathenow, & H. Starkey (Eds.), *Teaching for citizenship in Europe*. Stoke-on-Trent, England: Trentham Books.
Pogge, T. (2008). *World poverty and human rights: Cosmopolitan responsibilities and reform*. Malden, MA: Polity.
Rapoport, A. (2008, November). *We cannot teach what we don't know: Indiana teachers talk about global citizenship education*. Paper presented at the annual meeting of the National Council for the Social Studies, Houston, TX.
Rapoport, A. (2009). Lonely business or mutual concern: The role of comparative education in the cosmopolitan citizenship debates. *Current Issues in Comparative Education*, 12(1). Retrieved from http://www.tc.edu/cice
Schweisfurth, M. (2006). Education for global citizenship: Teacher agency and curricular structure in Ontario schools. *Educational Review*, 58(1), 41–50.
Segrera, F. (2010). *Trends and innovations in higher education reform: Worldwide, Latin America and in the Caribbean* (CSHE.12.10). Berkeley: University of California, Center for Studies in Higher Education.
Spring, J. (2008). Research on globalization and education. *Review of Educational Research*, 78(2), 330–363.
Subedi, B. (Ed). (2005). *Critical global perspectives: Rethinking knowledge about global societies*. Charlotte, NC: Information Age.
te Velde, D. W. (2008, August). *Globalisation and education: What do the trade, investment and migration literatures tell us?* London: Overseas Development Institute. Retrieved from http://www.odi.org.uk/resources/docs/2484.pdf
Tully, J. (2008). Two meanings of global citizenship: Modern and diverse. In M. Peters, H. Blee, & A. Britton (Eds.), *Global citizenship education: Philosophy, theory, and pedagogy*. Rotterdam, the Netherlands: Sense.

VanSledright, B. (2008). Narratives of nation–state, historical knowledge, and school history education. *Review of Research in Education, 32*(1), 109–146.
Westheimer, J., & Kahne, J. (2004a, April). *Educating the good citizen: Political choices and pedagogical goals*. PSOnline. Retrieved from http://www.apsanet.org
Westheimer, J., & Kahne, J. (2004b). What kind of citizen? The politics of educating for democracy. *American Educational Research Journal, 41*(2), 1–30.
Zong, G. (2009). Developing preservice teachers' global understanding through computer-mediated communication technology. *Teaching and Teacher Education, 25*(5), 617–625. doi:10.1016/j.tate.2008.09.016
Zong, G., Wilson, A. H., & Quashiga, A. Y. (2008). Global education. In L. S. Levstik & C. A. Tyson (Eds.), *Handbook of research in social studies education* (pp. 197–216). New York: Routledge.

CHAPTER SIX

Grounding Globalization: Theory, Communication, and Service-Learning

Ozum Ucok-Sayrak
Erik Garrett

As Tomlinson (1999, p. 1) states in *Globalization and Culture*, "Both globalization and culture are concepts of the highest order of generality and notoriously contested in their meanings." Along these lines, Pieterse (as cited in Parker 2005, p. 5) notes that "there are almost as many conceptualizations of globalization as there are disciplines in the social sciences." Furthermore, "teachers and scholars in disciplines such as management, marketing, finance, accounting, and economics also use the term 'globalization' to mean different things" (Parker, 2005, p. 5).

There are different views regarding the history of globalization; some argue that it is a recent Western, and particularly American, project, "a process that has emanated from and been greatly shaped by Western or American hegemony" (Robertson & White, 2003, p. 9). As Robertson and White (2003, p. 9) explain further, some regard globalization to be a very long historical process. To illustrate, they state that the United States came into existence at the end of the eighteenth century and is itself a product of the globalization process along with other nation-states of the Americas.

Given the variety of viewpoints on globalization and its history, our first argument about exploring globalization is that scholars need to be cautious with reductionistic approaches in making sense of a multifaceted and complex process such as globalization. In our experience of teaching college-level intercultural communication courses, we explicitly make it a point to acknowledge the diversity of perspectives and the evolving meanings constructed around globalization. One of the pedagogical challenges we face in discussions of globalization in the classroom is to make the abstract notion of globalization concrete for students. Therefore, our second argument suggests that the teaching of globalization can become more concrete through attending to local experiences in one's immediate community that give insight into global experiences. Offering students an opportunity to taste local experiences of alterity provides them an experience of something other than themselves. Thus, in the local, practical lessons of the global can emerge.

A specific pedagogical technique that can address the challenge of teaching globalization through the context of local experience is service-learning. Service-learning can help students reflect on globalization when the local experiences of helping their neighbors highlight the diversity that exists in their own community. Powerful local experiences can help students transcend the boundaries of sameness and could open them up to the fact that not everyone sees the world as they do. It is hoped that students can become aware that their view of the world is one constructed story among a community of others. Taking inspiration from scholars such as Agbaria (2011), we start this chapter by highlighting the importance of attending to the issue of globalization as symbolic discourse. That is, we acknowledge the symbolic construction of the meaning of globalization in making sense of this multifaceted phenomenon and connect it to global education, specifically social studies and intercultural communication. In the next section, we provide a survey on the literature of globalization as symbolic discourse, not only highlighting connectedness but also disconnectedness. Then, we provide a discussion of global civics and ethics in the context of rhetoric, warning about the potential for violence and conflict as part of the globalization process. Finally, in preparing our students to be global citizens we offer service-learning as a pedagogical practice, bringing the local in conversation with the global.

Literature Review

As our introduction suggests, there is a wide variety of definitions, descriptions, and perspectives on globalization across various fields of study—not only limited to social sciences—and, the meaning of globalization refers to different processes depending on the purpose of inquiry. To add a deeper—and more complicated—discussion to the above, Bartelson (2000) challenges the wide acceptance of globalization as a fact, "as a process of change taking place 'out there,'" despite the lack of agreement and ambiguity about what globalization is. "While there is no agreement about what globalization is, the entire discourse on globalization is founded on a quite solid agreement that globalization is" (Bartelson, 2000, p. 180). He further states that the factuality of globalization is worth investigating, "especially since it might be argued that this fact is partly constitutive of what globalization is about: nothing changes the world like the collective belief that it is changing, albeit rarely in directions desired by the believers" (Bartelson, 2000, pp. 180–181).

Bartelson's perspective is important to keep in mind given the rising popularity of the concept of globalization in recent years. The concept of "global teens" exemplifies this point of view. Parker (2005, p. 6) uses the example of

Grounding Globalization 69

the "middle-class phenomenon of 'global teens' who share interests, fashions, and musical tastes worldwide." She further goes on to state that the emergence of the global teen as a cultural phenomenon is interconnected with the existence of a global telecommunications infrastructure, and businesses and industries that are able to produce global brands for the global teen (which is connected to the global political environment that facilitates trade, and a global economy and environment that support the trade).

Similar to Parker, Hassan and Katsanis (1991) present the "global teen" as a fixed social/cultural reality, which leads to a reification of the concept.[1]

> Global teens from New York, Tokyo, Hong Kong, to those from Paris, London, and Seoul are sharing memorable experiences [through television, international education, and frequent travel] which are reflected in their consumption behaviour. . . . The "teenage culture" on a global scale shares a youthful lifestyle that values growth and learning with appreciation for future trends, fashion and music. (Hassan & Katsanis, 1991, p. 21)

The interconnectedness of the various global spheres is important to highlight. However, the uncritical acceptance of the "global teen" concept as a factual phenomenon reduces the global reality of all youth into a middle-class Western consumerist norm. Along these lines, Kjeldgaard and Askegaard (2006, p. 246) argue that "the myth of a global youth segment is a direct product of marketers' own ideologically framed cultural constructions via advertisements, practitioner-oriented literature, and various other forms of cultural production." The authors state that the global youth segment as a "transnational market ideology" emerged in relation to the development of global marketing.

> For marketers and consumers alike, such signs of a global youth culture are all too readily treated as obvious evidence of a homogenized group of consumers. In the Barthesian tradition, myth is exactly such a naturalization of a social set of signs (Barthes, 1957). The myth of a global youth hence constitutes an ideological explanatory framework for practices observed in social reality. (Kjeldgaard & Askegaard, 2006, p. 231)

Thus, one needs to be critical and cautious in reading news and "facts" about "global teens" or any other concept related to globalization including the term itself.

It is important to hear what some scholars (Agbaria, 2011; Fiss & Hirsh, 2005) say about globalization as a symbolic discourse that complements the majority of studies that focus on globalization as a structural process. Agbaria (2011, p. 69) points to the studies on globalization in educational research that examine the influence of globalization on education and discuss reforms in educational systems within the context of globalization without offering much clarity "as to what kind of globalization students around the world are being prepared for." He goes on to add, "It is not fully lucid what kind of globalization exists 'out there' that social studies and educators in related disciplines are encouraged to believe in and to circulate as objective truth" (p. 69).

Thus, Agbaria (2011) highlights the limited insights offered by academic studies on the conceptualization and framing of globalization in specific discourse sites and communities in education. Examining the debates on the history and definition of globalization between 1990 and 2005 in two major social studies education journals in the United States (*Social Education* and *The Social Studies*), Agbaria (2011) states, "By keeping an eye on the language we use to discuss globalization, we will be better able to provide a critical view on the certainty of many globalization images and understandings" (p. 77). Furthermore, the democratization of the social studies education discourse community on globalization and global education involves the inclusion of "alternative narratives, skeptical perspectives and critical voices" and an exploration of the "discursive forces at work in creating how globalization is imagined" rather than approaching globalization as an objective fact (Agbaria, 2011, p. 78).

Highlighting the symbolic construction of globalization, however, is not to ignore the material dimension of the globalization process, but to promote a critical and reflexive engagement in making sense of a multifaceted phenomenon. Tomlinson (1999, p. 2) discusses globalization as a process of "complex 'connectivity'" that "refers to the rapidly developing and ever-densening network of interconnections and interdependencies that characterize modern social life." These networks include a growing intensity of transnational flows in the form of people, commodities, ideas, technologies, capital, social movements, diseases, fashions, beliefs, images, pollution, and so on that characterize globalization as "an empirical condition of the modern world" (Tomlinson, 1999, pp. 1–2). Tomlinson (1999) explores what it culturally means to live in a more globally connected world, and the significant relation between connectivity and proximity. He states that the ease and availability of rapid travel between distant places as well as the use of media technologies that allow for the transmission of images and information around the world have affected people's sense of distance. People now perceive physically distant places and people as being close and accessible (either physically, or representationally through the media), which is also expressed through globalization discourses such as the "shrinking world" or "global village" (McLuhan, 1992). This changing sense of proximity might also imply an emerging intimacy, a "global

Grounding Globalization 71

closeness"; however, as Tomlinson (1999) articulately illustrates through a phenomenological account of airline travel, accessibility, and routinization, this does not necessarily correspond to cultures coming closer in the sense of intimacy.

> After a few hours of this enclosed time-journey we arrive, clear customs, walk out of the terminal building in the same clothes in which we boarded (tangible attachments to our not-so-distant home) into a strange environment, a different climate, probably a different language, certainly a different cultural tempo. What sort of "proximity" does such a process involve? How, precisely, has the connectivity provided by air travel brought us closer? . . . For the space we traverse in these journeys through the routine sequence of "cabin time" is not just physical distance but the social-cultural distance (Saudi Arabia = Islam = no alcohol) that "real" material space preserves. The connectivity of air travel thus poses for us sharply the question of the overcoming of social-cultural distance. (Tomlinson, 1999, p. 5)

Shome and Hedge (2002, p. 174) support Tomlinson's discussion and illustration above by stating that globalization refers to "an entrenched pattern of worldwide connectedness (Held & McGrew, 2000) and disconnectedness, (an issue that seems to receive less attention)". The discussion of increasing connectivity and proximity/intimacy brings up a question: Living in a more connected world, are we "inevitably being drawn together, for good or ill, into a single global culture" (Tomlinson, 2006, p. 2)?

Communication Studies and Globalization

The contentious understanding of globalization has been further supported in the discipline of communication. Globalization is of central importance in both intercultural communication and rhetoric. In the intercultural communication literature, scholars such as Rodriguez and Chawla (2010) have taken up Friedman's (1999) and Chua's (2003) understanding of globalization to highlight the connection between globalization and violence in communicating betwixt and between persons and cultures. Friedman is sympathetic to globalization. In *The Lexus and the Olive Tree: Understanding Globalization*, he puts forth the idea of the "Golden Arches Theory of Conflict" (1999). This theory states that one can measure the "progress" of a particular culture by whether it has a McDonald's. For Friedman, having a McDonald's represents the capacity for that culture to have a middle class, and this indicates a stability that decreases the likelihood for conflict and violence. However, according to Chua (2003), the exact opposite is true. The McDonald's is a representation of a huge culture change and is a constant reminder to the host culture of something lost and that increases anxiety over change. Whether positive or negative, globalization is a real lived experience that must be dealt with when speaking across cultures.

The other area where globalization is of central importance to the discipline of communication is in the connection to the ancient tradition of rhetoric. From the pre-Socratics to modern scholars we know of rhetoric as "the art of persuasion." In this broad understanding of rhetoric, its content concerns discourse about any cultural object. The classical rhetoricians were concerned about training the citizen to take his place in the public sphere. This emphasis on the good citizen speaking well (Quintilian, 1987) is also the progenitor of modern civics. The debate over globalization has problematized how we are to interpret the rhetorical tradition itself. As a matter of fact, the question about the influence of the local and the global in culture that globalization scholars are discussing is also challenging the way rhetoricians are talking about their discipline. For example, Gaonkar (1997) sees a problem when rhetoric is interpreted from a global position as talking across any cultural boundary. According to Gaonkar, this purely open interpretation detaches rhetoric from its "humanist" roots and tradition. If we loosely interpret rhetoric in the context of a global text, the question becomes, do we lose its particular ground? So the point for Gaonkar is that rhetoric itself needs to avoid the Scylla and Charybdis of globalization and particularity. For when rhetoric is construed under the context of globalization it loses the local voice that is so fundamental to the study of the discourse of persuasion. Other communication scholars, such as Dutta-Bergman (2005), object to the rhetorical traditionalists use of "civil society" as a hegemonic way of controlling and taming the globalized other. Rhetoric itself becomes a tool (or weapon) that "perpetuates top-down relationships among groups of nations." (Taylor, 2009, p. 77) The response to the critique about the hegemony of a globalized rhetoric has been to emphasize the pragmatic side of rhetorical theory (Leff, 1993). Rhetoric is not merely theory, but also praxis. Therefore, rhetoric as situated practice is mindful of the myriad of differences from which the speaker speaks to a wide array of audiences.

The question of violence in globalization as discussed earlier as well as the discussion of a hegemonic globalized rhetoric in controlling and taming the globalized other naturally leads to engagement with the ethical ramifications of the globalization process. As Sadri and Flammia (2011, p. 253) state, "As global citizens, perhaps one of the greatest challenges we face is developing a code of ethical behavior that can address the many diverse issues we are likely to encounter when communicating across cultures." In the field of communication, the question of the ethical goes back to our ancient rhetorical roots. This is because rhetoric is fundamentally a project of intersubjective discourse. We always speak to someone "Other" than ourself. The famous philosopher of ethics Emmanuel Levinas (1969) has pointed out that the face of the Other represents an ethical commandment. Levinas argues that in the face-to-face relationship we are responsible for the Other because we are our "brother's keeper." Ultimately, he claims "ethics as the first philosophy." (Levinas, 1969, p. 304)

Service-Learning and Global Education

The practical anchor that rhetoric as situated practice offers to the challenges of globalization can also be found in the pedagogical practice of service-learning. Service-learning as a form of experiential learning can trace its roots back to this rhetorical emphasis on praxis. However, in the modern education theory literature on service-learning the origins of the technique are grounded in American pragmatism and the works of John Dewey. According to Giles and Eyler (1994), Dewey provides the "theoretical roots" of service-learning. Dewey's pragmatism emphasized the practical aspect that theory provides, especially in the context of education. Service-learning must also contextualize the globalization debate into its engaged theoretical perspective. Despite the ease of travel and technology, we cannot always transport the classroom around the world in a manner that engages in service in a meaningful manner. Granted, some get-away and summer study abroad classes have adopted this methodology. Yet we must ask: How does the service-learning instructor, teaching in the context of a local school neighborhood, help expand the horizons of students to pay attention to questions of globalization and the global neighbor? To do this we suggest an interpretation of service-learning as teaching the global through local practices. It is our attempt to raise global awareness through local service that we want to make explicit in the remainder of this chapter.

The question of globalization is ultimately one of context. One always communicates from somewhere via a phenomenologically embedded position. Part of being a "good citizen" is recognizing the contingent ground upon which we walk in relation to Others. At Duquesne University this enriched notion of citizenship is pedagogically instilled in our University Core. One of the courses that meet the Global Diversity requirement of our Core is the Exploring Intercultural Communication class. This class meets the Core's theme of being attentive to the challenges of globalization. So, rather than teaching the course as a "tour guide" through foreign and exotic cultures, we instead have the class attend to the notion that culture is something that is active and everywhere. We teach global diversity from a narrative of service that puts the local in conversation with the global. In Exploring Intercultural Communication, the students enter into the global through local narratives. Our students explore the city of Pittsburgh to listen to and document the local stories of urban communities in need. They go to the city's most economically disadvantaged neighborhoods and conduct both formal and informal interviews while engaging in projects such as urban gardening and neighborhood clean-ups, or attending local community meetings. Students work side by side with the community. Back in the classroom, we talk about how the stories we heard, along with our own experience and observations, and share important human themes that transcend the local community.

For example, one student who volunteered in a community event sponsored by the Homeless Children's Education Fund, "Stand Up for Homeless Children: A Performance Art Installation," described that she was given a picture of a homeless child and was instructed to hold it in front of her face along with other volunteers in a busy public space in the city. They were to stand and remain silent for fifteen minutes. After fifteen minutes, all of the volunteers simultaneously said, "Listen!" ten times from behind their picture in order to make their message heard to those in their surroundings. In her service-learning paper, she reflected as follows:

> On my way to this event, I found myself feeling annoyed and agitated because I didn't truly understand what the event represented and all I knew was that I was going to be standing in the freezing cold weather, holding a picture. It seemed pretty ridiculous to me at the time, however, all of that changed as the event unfolded. When I was given my picture to hold, and started learning of the hundreds of homeless children in Allegheny County alone, I began to feel sincere sympathy for these children. As I was standing in the cold, with everything around me completely silent, I thought of how these children are forced to live in this weather all day, every day and here I was complaining about doing it for fifteen minutes. It really opened my eyes to how fortunate I was to have a home and food to eat whenever I wanted, and how I should be much more considerate and open-minded towards events such as this one. (Student communication, 2011)

This student's narrative powerfully illustrates the embodied learning that lies at the core of service-learning. Service-learning has been defined as "experiential learning" (Giles & Eyler, 1994). The various service activities of the students, such as the one above, allow for "hands-on experience" and reflection of different issues ranging from homelessness to racial conflict to urban blight. The direct experience of another human being's suffering, though brief, allowed this student to see the world from the Other's perspective for a moment, which further allowed her to realize her own standpoint in life. During class discussion of the service-learning experiences, another student who also volunteered in the same community event shared a similar reflection and stated that it allowed her to place herself in the shoes of the Other. This experience actually led to a powerful realization that she "might not actually be able to understand what these people are going through," because she never had limited opportunities in her life and always had support from her parents. Then she added, "But I can try and do my best to help them in any way I can" (student communication, 2011). From the narratives we shared above, there seems to be a natural extension from the experience of attending to the local, the particular, with care to a larger, global connection of the human experience with a sense of an ethical responsibility for the Other. Arnett (2008, p. 70) states that:

> the human face is a provincial signpost that redirects us to an ethical echo a priori to the Western construct of Being and, in communication ethics, prior to the construction of an autonomous ethical self. The visage of another takes us from sight to attentive listening to an ongoing quiet mantra: 'I am my brother's keeper.'

Attending to the "first human home of provinciality" (Arnett, 2008, p. 70), that is, the human face, by standing up for homeless children, the two students not only expressed a connection but also an ethical responsibility to reach out and attend to the Other. Service-learning activities seek to cultivate this human connection and a sense of responsibility that emerges out of such connection through attending to the particular ground, and "a particular sense of provinciality," that lies at the heart of "the phenomenological reality of human sense-making" (Arnett, 2008, p. 74).

Conclusion

One of the claims of this chapter is that globalization must not be treated in a reductionistic way. Globalization is a complex, fluid, and multidimensional concept. Our second argument is that through attending to the empathic potential of local experiences of difference, we can awaken a global understanding. We acknowledge that even in our local community, we cannot take for granted that everyone shares the same beliefs, values, norms, and way of life. Attending to diverse local experiences can give insight into global experiences through embodied learning.

We support these arguments first through a review of the literature on globalization as symbolic discourse. Then, we provide a discussion of civics and ethics as it arises from the rhetorical tradition. Finally, we offer service-learning as a potential pedagogical practice to bring the local in conversation with the global.

Our contribution in this piece is that discussions of globalization must ultimately be grounded somewhere. The ground from where we speak matters. That is why service-learning is not a merely provincial pedagogical technique. In listening to the dialect of the local story we can hear the tale of globalization. Helping the neighbor at home can connect us to the neighbor across the globe. If we are to break away from the "structural violence" (Demenchonok & Peterson, 2009) inherent in global capitalist systems we must reinvigorate the meaning of being a neighbor. We offer the "experiential learning" of the service-learning pedagogical practice as one way to help students learn to speak across global distances.

Note

1. An important critique of Parker's "global teen" phenomenon is that it is obviously Western, Eurocentric, white, and consumerist. This phenomenon is problematic because it insidiously takes these dominant cultural practices and institutionalizes them in a normative frame that washes over and ignores the vast intercultural difference in which teens around the world practice their culture. So, while the "global teen" example may work in certain Western cultures, we need to question its applicability to places such as India, Gabon, Trinidad, and Ecuador.

References

Agbaria, A. (2011). Debating globalization in social studies education: Approaching globalization historically and discursively. *Intercultural Education*, 22(1), 69–82.

Arnett, R. (2008). Provinciality and the face of the other: Levinas on communication ethics, terrorism—otherwise than originative agency. In K. G. Roberts & R. Arnett (Eds.), *Communication ethics: Between cosmopolitanism and provinciality*. New York: Lang.

Bartelson, J. (2000). Three concepts of globalization. *International Sociology*, 15(2), 180–196.

Chua, A. (2003). *World on fire: How exporting free market democracy breeds ethnic hatred and global instability*. New York: Doubleday.

Demenchonok, E., & Peterson, R. (2009). Globalization and violence: The challenge to ethics. *American Journal of Economics and Sociology*, 68(1), 51–76.

Dutta-Bergman, M. (2005). Civil society and public relations: Not so civil after all. *Journal of Public Relations Research*, 17(3), 267–289.

Fiss, P. C., & Hirsch, P. M. (2005). The discourse of globalization: Framing and sensemaking of an emerging concept. *American Sociological Review*, 70, 29–52.

Friedman, T. L. (1999). *The Lexus and the olive tree: Understanding globalization*. New York: Farrar, Straus and Giroux.

Gaonkar, D. (1997). The idea of rhetoric in the rhetoric of science. In A. Gross & W. Keith (Eds.), *Rhetorical hermeneutics: Invention and interpretation in the age of science*. Albany, NY: SUNY Press.

Giles, C., & Eyler, J. (1994). The theoretical roots of service-learning in John Dewey: Toward a theory of service-learning. *Michigan Journal of Community Service Learning*, 1.

Hassan, S. S., & Katsanis, L. P. (1991). Identification of global consumer segments: A behavioral framework. *Journal of International Consumer Marketing*, 3(2), 11-28.

Held, D., & McGrew, A. (2000). The great globalization debate: An introduction. In D. Held & A. McGrew (Eds.), *The global transformations reader: An introduction to the globalization debate* (pp. 1-46). Cambridge: Polity.

Kjeldgaard, D., & Askegaard, S. (2006). The globalization of youth culture: The global youth segment as structures of common difference. *Journal of Consumer Research*, 33, 213-247.

Leff, M. (1993). The idea of rhetoric as interpretative practice: A humanist's response to Gaonkar. *Southern Communication Journal*, 58(4), 296-300.

Levinas, E. (1969). *Totality and infinity: An essay on exteriority*. Pittsburgh, PA: Duquesne University Press.

McLuhan, M. (1992). *The global village: Transformations in world life and media in the 21st century.* New York: Oxford University Press.

Parker, B. (2005). *Introduction to globalization and business: Relationships and responsibilities.* Thousand Oaks, CA: Sage.

Quintilian. (1987). *On the teaching of speaking and writing* (J. J. Murphy, Trans.). Carbondale: Southern Illinois University Press (original work appeared in 95 CE).

Roberts, K. G., & Arnett, R. (Eds). (2008). *Communication ethics: Between cosmopolitanism and provinciality*. New York: Peter Lang.

Robertson, R., & White, K. E. (2003). *Globalization: Critical concepts in sociology.* (Vol. 1). New York: Routledge.

Rodriguez, A., & Chawla, D. (2010). *Intercultural communication: An ecological approach.* Dubuque, IA: Kendall Hunt.

Sadri, H. A., & Flammia, M. (2011). *Intercultural communication: A new approach to international relations and global challenges.* New York: Continuum.

Shome, R., & Hedge, R. (2002). Culture, communication, and the challenge of globalization. *Critical Studies in Media Communication*, 19(2), 172-189.

Taylor, M. (2009) Civil society as a rhetorical public relations process. In Heath, R., E. Toth, & D. Waymer, (Eds.), *Rhetorical and critical approaches to public relations II* (pp. 76-91). New York: Routledge.

Tomlinson, J. (1999). *Globalization and culture.* Chicago: University of Chicago Press.

Tomlinson, J. (2006). *Globalization and culture.* Retrieved from The University of Nottingham Ningbo China Public Lectures series.

CHAPTER SEVEN

Global Classrooms: A Contextualized Global Education

Cameron White

> Global educators share certain characteristic instructional strategies: they confront stereotypes and exotica and resist simplification of other cultures and global issues; foster the habit of examining multiple perspectives; teach about power, discrimination, and injustice; and provide cross-cultural experiential learning.
>
> Merry M. Merryfield, 2008

Introduction

Growing up today is very different from even a decade ago. Today the world truly is a smaller place and we really are part of a global neighborhood. People are instantaneously connected to international events through media, technology, trade, and global issues such as conflict, climate, and socio-economics. Borders do not mean the same thing as they did just a few years ago. A "globalized" world necessitates international connections, thus challenging traditional conceptions of nationhood and exceptionalism.

This chapter analyzes the concept of global education and suggests that we need to improve our contextualization of the issues and investigation of global connections. The chapter also suggests that a variety of popular culture texts and more student-centered approaches should be the focus. In addition, the concept of Global Classrooms is described as an alternative approach to meaningfully allowing students to analyze and engage in global education through research, problem solving, debate, and role-playing.

Teaching and learning must include education for a global perspective so that students might also become responsible "active" citizens of the world (Chapin, 2003; Diaz, Massialas, & Xanthopoulos, 1999; Tucker & Evans, 1996). A critical component of education in general, and social studies specifically, is to promote an understanding of diversity at home and abroad: "Integrating global realities within an existing school curriculum meets the needs of an ever-changing, ethnically diverse, increasingly interdependent, international community" (Tucker & Evans, 1996, p. 189). World citizenship requires a global education.

Global education efforts must begin with an attempt to understand globalization. Globalization can be defined as "the intensification of worldwide social relations which link distant localities in such a way that local happenings are shaped by events occurring many miles away and vice versa" (McLaren, 1995, p. 180). Diaz and colleagues (1999, pp. 37–38) state that globalization

> refers to the compression of the world and to the intensification of the consciousness of the world as a whole. This process is ongoing and all of us, young and old, Westerners and non-Westerners, are inescapably involved in it. The compression of the world is real. People witness it in their daily lives, in the foods they eat, in the TV programs they watch, in the cars they drive, in the dresses and costumes, in the people they choose to govern them, and so on.

Clearly, globalization is increasingly influential in all aspects of life. Therefore, understanding it through global education is imperative. Schools must provide opportunities for children to "develop the appropriate cognitive skills to understand and explain the globalization process and to critically analyze its impact on their lives and the lives of people around them" (Diaz et al., 1999, p. 38). Above all, according to Diaz and colleagues (1999), students need to know how to affect the global system as world citizens and as advocates of a well-grounded position or point of view. This suggests that students must acquire both a new knowledge base and a skill set. Many of the subjects associated with social studies might offer an appropriate space for global education—but other disciplines are ripe for integration as well. At its core, global education is really about analyzing the links between cultures and people (Chapin, 2003); it must be better integrated in all classrooms.

Rethinking teaching and learning in these ways could provide the opportunity to deepen our understanding and appreciation of others in the world, something essential to our roles and responsibilities as global citizens. Given the global interconnectedness of the world today, the global context must be present. According to Merryfield (2001), students must develop a global perspective that will emphasize cross-cultural experiential learning and stress commonalities in cultures that transcend diversity. One culture that already seems to transcend world barriers is popular culture.

Increased globalization presents many challenges for societies and the institution of education has a responsibility for addressing these "globalized" issues. Education in general should play a strong role and is enhanced through internationalizing partnerships and projects. In addition, cultural competence, collaboration skills, and an appreciation of global connections can be facilitated through cross-cultural experiences in teacher education, thus translating to the classroom. Many schools, colleges, and universities are recognizing the need for global competence and promoting understanding among cultures (Dan-xia, 2008). In addition, linking multicultural education and global issues is facilitated through meaningful international education projects (Wells, 2008). James (2005) suggests that internationalizing education develops a sense of interconnectedness, empathy, and tolerance, all of which are much needed in today's world. Education programs have much work to do to accomplish these goals. A way to move forward is to share ideas and engage in collaborative internationalizing of the curriculum.

Contextualizing the Global through Popular Culture

Our students need alternative texts to enable teaching and learning to come more alive—popular culture texts are just such tools. Today, popular culture is more specifically associated with commercial culture: movies, television, music, advertising, toys, photography, games, the Internet, and so on. Popular culture is also one of the United States' most lucrative exports. Many critique American popular culture as a form of cultural imperialism tied to unbridled capitalism. While this issue should definitely be included in any integration of popular culture, cultural sharing goes many ways.

The export of popular culture is not a new phenomenon. Elements of popular culture have always spread beyond nationalist borders. Today, satellite television, multinational media corporations, and the Internet provide unprecedented opportunities for the spread of popular culture. Students in the United States have access to media that allows them to experience cultures from around the world. Conversely, many of these culture "industries" began with U.S. exports, so the world's access to American popular culture is also burgeoning. Given the global cross-cultural nature of contemporary popular culture, it is sometimes difficult to determine origin. This globalization of popular culture has resulted in what Kellner (1995) refers to as a new "global popular."

The existence of the global popular means that we might easily find a common frame of reference or topic when asking U.S. students to consider "others" in the world. Integrating themes and issues connected to Global Classrooms or Model United Nations by using popular culture can only enhance student interest and connections. Two themes that have great possibility while integrating popular culture include investigating global issues and differing cultures by looking deeply at media representations of such issues (news, technology, film, music, art). Inquiry into how various issues are represented in media, advertising, and other forms of popular culture can enhance deeper understandings and connections for our students.

Schools in the United States must provide opportunities for students to learn about the world: who people are, what they do, and how they live. Students must learn how to get along with all people within the United States and around the world as responsible citizens of both. Education for civic competence, for responsible national and world citizenship, falls within the domain of social studies instruction and learning. We must rethink teaching and learning to enable these ideas. Integrating music, movies, art, and literature focusing on global issues or celebrating global culture offers a great opportunity.

Global Classrooms: An Integrated Model

The unique quality of the Global Classrooms (GC) program when compared to the traditional Model UN concept is found in its integrated issues and problem-based curriculum, skills development, and support provided throughout an entire school year. Model UN, on the other hand, is most often a one-time simulation of the United Nations. In addition, while Model UN is often available only to more privileged schools and communities, Global Classrooms is designed for schools and students who don't always have access to great resources.

Global Classrooms is a program that was initiated by the United Nations Association some twelve years ago to facilitate the integration of Model UN and Global Classrooms skills-based curriculum into schools that had not experienced traditional Model UN. The idea is to support each portion of the program in more socio-economically deprived schools, generally in urban areas. Primary goals of Global Classrooms include developing global/international knowledge based on global issues and countries, developing a variety of life skills leading to active global citizenship, and developing dispositions focusing on awareness and appreciation of multiple perspectives.

Global Classrooms has a constructivist-oriented curriculum based on several themes, including peacekeeping, human rights, sustainable development, and globalization. Within each curriculum framework there are lessons that focus on developing a UN knowledge base and a general awareness of global issues. In addition, each lesson focuses on skills development, including research, writing, collaboration, debate, and general public speaking. Dispositions encouraged through the curriculum include tolerance, awareness, collaborative demeanor, openness, and appreciation of difference.

A variety of thematic simulations are included to prepare students for the role-playing experience of Model UN (MUN). In addition, the curriculum is generic enough to encourage contextual adaptation to localize issues and even include less controversial simulations such as debating which type of candy is best, or the best fast-food restaurants. Resources, lessons, and links are also included to support teaching and learning.

The program was initially supported through large grants that paid for the curriculum, professional development, the MUN conference, and support staff. Local United Nations Associations (UNA) generally served as the liaison between the UNA-USA in New York, local schools, and a Global Classrooms consultant. At present, most local Global Classrooms programs receive minimal funding from UNA-USA, although curriculum is still free. Staff members are paid minimally, if at all, and at present there is a charge to participate in the conference. The charge covers the venue, lunches, some staff per diem, and basic conference support. Regardless of the sharp drop in funding support, the program is sustainable and has grown to include several U.S. and international cities.

GC Houston: A Case Study

Global Classrooms Houston is one of the original programs supported by the UNA-USA and has been in existence since fall 2001. Although it was initially mandated that GC Houston be available only to Houston ISD, it expanded into Aldine ISD first and is now located in several local districts and charter schools. Generally, over three thousand students engage in some form of Global Classrooms during the school year, with five hundred students participating in the culminating Global Classrooms; a Model United Nations Conference is held at the University of Houston each spring.

Houston was initially unique in that it was one of the first cities to have both a middle school and high school Global Classrooms project, as local schools recognized a need to integrate the program in sixth-grade world cultures classes that focused on the twentieth century as well as various high school subjects. Houston schools also integrated the curriculum in a variety of ways—within the specific social studies class itself, in clubs, and in stand-alone elective courses. Even within each of these, individual teachers were able to adapt the curriculum in a variety of ways—as a current events/issues-based integration, as a way to develop life skills in clubs such as debate or speech, or as a semester- or year-long Global Classrooms course integrating various components of the curriculum.

GC Houston holds several professional development sessions each year, focusing on introduction to Global Classrooms and Model UN, curriculum integration, conducting Model UN simulations, and conference preparation. Fall is typically used for curriculum integration, mock debates, and knowledge and skills development for the students. Spring is dedicated to conference preparation. Schools are assigned countries and topics depending on requests and numbers of students attending the conference. During the culminating conference, six committees are usually represented at both the middle school and high school levels, including the Security Council and other committees of high interest such as UNICEF.

Students are assigned countries, research their countries, are assigned committees, and research their committees. Then they are assigned topics, research the topics, and articulate what their country's position is. Students develop a position paper to assist with debate during the conference. Students engage in mock debate to prepare for the conference, usually focusing on localized issues. Ultimately students are asked to role-play UN ambassadors from their assigned countries. GC Houston has adapted the curriculum in a number of ways, but localized themes include energy, transportation, pollution, development, space issues, trade, education, and socio-economic issues—focusing on contextualized issues in Houston.

Despite the general success of Global Classrooms Houston, the program has been challenged in local schools and districts and in Texas for addressing controversial issues, allowing for multiple perspectives, and not fitting into the Texas Essential Knowledge and Skills requirements. Much of this is addressed by inviting critics to visit classrooms, observe student activities, and attend the culminating Model UN conference. Stakeholders generally agree that Global Classrooms provides a contextualized global education missing from the current traditional curriculum. The largest ongoing issue is time, as teachers are increasingly subjected to scripted curricula, benchmark testing, and formalized standardized testing, and to narrow perceptions regarding achievement and accountability.

Extensions

Numerous resources are available that enable improved global education in addition to Global Classrooms and Model United Nations. For example, in addition to the GC Curriculum, the UNA-USA recommends the integration of "Educating for Global Competence," published by the Asia Society (http://www.ccsso.org/Resources/Publications/Educating_for_Global_Competence.html). What is especially appropriate is that as with Global Classrooms, there are four global competencies promoted, taught, and implemented, including investigating the world, recognizing perspectives, communicating ideas, and taking action. According to the publication, contemporary societies are marked by new global trends—economic, cultural, technological, and environmental shifts that are part of a rapid and uneven wave of globalization. The growing global interdependence that characterizes our time calls for a generation of individuals who can engage in effective global problem solving and participate simultaneously in local, national, and global civic life. Put simply, preparing our students to participate fully in today's and tomorrow's world demands that we nurture their global competence. Global competence is the capacity and disposition to understand and act on issues of global significance. (Asia Society, 2011, http://www.ccsso.org/Resources/Publications/Educating_for_Global_Competence.html)

Facing the Future (http://www.facingthefuture.org/) is another of many examples that provides lessons and links as extensions for developing global education. Free downloadable curricula focus on themes such as climate change, water, science, civics, and consumerism. Again, the conceptual framework centers on critical thinking, global perspectives, and informed actions, as stated on the Web site.

Another example comes from Global-Ed.org (www.global-ed.org/curriculum-guide.doc) and resonates well with the previous examples and with Global Classrooms. This program encourages a thematic approach focusing on development, the environment, human rights, and peace—all through an education framework. Goals of the program include developing respect, valuing and celebrating other cultures, learning about developing countries and their issues in a positive way, becoming socially and environmentally responsible, gaining a positive outlook on their role in making the world a more peaceful and just place, and clarifying the connections to real life.

Conclusions

Students today live in a world made up of many texts; it is essential that they develop multiple literacies that will allow them to read the signs, symbols, and images (texts) of that world. Schools must provide students with "new operational and cultural 'knowledge' in order to acquire new languages that provide access to new forms of work, civic and private practices in their everyday lives" (Lankshear & Knobel, 2003, p. 11). Our classes could be the perfect place to excite students. We can introduce them to meaningful knowledge and issues, debates regarding globalization, and relevant problem-based global education, all of which can provide the context for developing the literacy skills necessary to interpret those issues and delve more deeply into them.

Reading global issues through context (simulations and role-playing that focus on life skills) offers students the opportunities to apply knowledge and develop skills in relevant ways, as Kincheloe (2005) recommends. Students learn where people live, their environments, and culture. Much the same can be said in learning about other societies through global issues.

Responsible global citizenship requires knowledge of "others" in the world, whoever they might be. It also requires the skills to understand and act in the best interest of the majority of the people. The knowledge base should include an understanding of who the other people in the world are, what they do, and where they are. The skill set should include inquiry and critical literacy/thinking skills. Curriculum and instruction must avoid the traditional coverage approach and provide students with opportunities to apply their knowledge in a meaningful context. Global Classrooms is a global framework that addresses these issues and can be used effectively in any classroom.

Links, Lessons, and Applications

United Nations Association
    http://www.unausa.org/Page.aspx?pid=220
Global Classrooms
    http://www.unausa.org/globalclassrooms
Model UN
    http://www.unausa.org/modelun
UN Cyberschoolbus
    http://www.un.org/cyberschoolbus/
United Nations
    http://www.un.org/en/
Center for Global Education
    http://globaled.us/
Global Education Resources
    http://www.nea.org/home/37409.htm
    http://resources.primarysource.org/globaleducation
    http://www.nais.org/sustainable/index.cfm?ItemNumber=146778
Teaching Globalization
    http://www.globalenvision.org/teachers
    http://essays.ssrc.org/sept11/essays/teaching_resource/tr_globalization.htm
    http://www.adifferentworld-unmondedifferent.org/
    Educating for Global Competence—the Asia Society
    http://www.ccsso.org/Resources/Publications/Educating_for_Global_Competence.html
Facing the Future
    http://www.facingthefuture.org/
    Global-Ed.org http://www.global-ed.org/curriculum-guide.doc

References

Chapin, J. (2003). *A practical guide to secondary social studies*. Boston: Pearson.

Dan-xia, C. (2008). A social distance study of American participants in a China study program. *US–China Education Review, 5*(9), 17–22.

Diaz, C., Massialas, B., & Xanthopoulos, J. (1999). *Global perspectives for educators*. Boston: Allyn & Bacon.

James, K. (2005). International education. *Journal of Research in International Education, 4,* 313–332.

Kellner, D. (1995). *Media culture: Cultural studies, identity and politics between the modern and the postmodern*. London: Routledge.

Kincheloe, J. (2005). *Getting beyond the facts*. New York: Lang.

Lankshear, C., & Knobel, M. (2003). *New literacies: Changing knowledge and classroom learning*. Buckingham, England: Open University.

McLaren, P. (1995). *Critical pedagogy and predatory culture: Oppositional politics in a postmodern era*. New York: Routledge.

Merryfield, M. (2001). Moving the center of global education: From imperial world views that divide the world to double consciousness, contrapuntal pedagogy, hybridity, and cross-cultural competence. In W. Stanley (Ed.), *Critical issues in social studies research for the 21st century* (pp. 179–207). Greenwich, CT: Information Age.

Merryfield, M. (2008). Scaffolding social studies for global awareness. *Social Education, 72*(7).

Tucker, J., & Evans, A. (1996). The challenge of a global age. In B. Massialas & R. Allen (Eds.), *Crucial issues in teaching social studies K–12* (pp. 181–218). Belmont CA: Wadsworth.

Wells, R. (2008). The global and multicultural: Opportunities, challenges, and suggestions for teacher education. *Multicultural Perspectives, 10*(3), 142–149.

CHAPTER EIGHT

Institutional Internationalization: The Undergraduate Experience

Linda B. Bennett

Higher education institutions incorporate internationalization/globalization into research, teaching, and service on their campuses (Knight, 1993), and one approach is the internationalization of the curriculum. Academic units across the country have internationalized undergraduate education through globalized curricula, e-learning, work/study abroad, international studies programs, global engagement certificates, foreign language study, or dual/joint degrees with partner overseas institutions. In 2006, 37 percent of institutions surveyed required an international or global focus course in general education, which was down 4 percent from 2001. At the same time, the percentage of institutions requiring a foreign language course in order to graduate for all or some students went from 53 percent to 45 percent (Green, Luu, & Burris, 2008). In the 2009-2010 Open Doors Report, the Institute of International Education (2011) found that 270,604 U.S. students studied abroad for credit, compared to 260,327 in 2008-2009. The 2008 *Mapping Internationalization on U.S. Campuses* survey found that 91 percent of institutions offer study abroad, which is a 26 percent increase since 2001, and 31 percent offer internships abroad, which is a 9 percent increase over the same time period (Green et al., 2008).

The undergraduate experiences at institutions across the United States provide students with academic opportunities to connect through technology and global engagement. Increasing usage of mobile devices, smartphone applications, and the Internet in the United States affects the experiences and expectations of undergraduates. According to the Pew Center's Internet & American Life Project August 2011 tracking survey, 94 percent of eighteen to twenty-nine-year-olds use the Internet and are more likely than the rest of the population to use social media and own Web-enhanced laptops or smartphones. The most popular smartphone applications in the United States are Facebook, Google Maps, and the Weather Channel ("The State of," 2010). The top three mobile applications for coursework are mathematical equations (52 percent), dictionaries/thesauruses (49 percent), and search tools (43 percent). Dahlstrom, de Boor, Grunwald, and Vockley (2011) believe that higher education can better integrate more of the technologies that students value into their learning experiences. There is the potential for educators to capitalize on the interests and abilities of students by incorporating networking into internationalization of the curricula. In addition, the use of technology in teaching and learning enhances the global knowledge economy (Arambewela, 2010) and prepares graduates to be national and global citizens (Qiang, 2003).

From one assignment in a specific course to a strategic plan for the infusion of global/international education and digital media into the infrastructure of a campus, faculty and administrators are designing curricula, courses, and programs for global learning. The descriptions of the instructional practices or programs from four campuses include samples of approaches used by students, faculty, and administrators to electronically capture information about international, intercultural, or global practices. This chapter presents examples of how four institutions offer undergraduate academic international/global experiences that integrate technology and use the institutional terminology and approach to globalization and/or internationalization.

The Global Awareness Program

The Global Awareness Program (GAP) at the University of Kansas (KU) was selected as an exemplary internationalization because it incorporates international experiences, foreign language, course work, and extracurricular activities. The program is one of the reasons KU was recognized by NAFSA with the Paul Simon Award for Campus Internationalization. The program has been in existence for over ten years and over 2,600 students have participated in it. The GAP portfolio guidelines and requirements can be found at http://www.international.ku.edu/gap/.

The Global Awareness Program is open to undergraduate students of any major who want to supplement their degree with a transcript certification for global engagement. Students complete two of the three global awareness components: an international experience, co-curricular activities and two semesters of modern foreign language or two international focus courses. If a student completes all three components, he or she is eligible to receive the GAP Certification with Distinction.

To fulfill the international experience, students earn KU credit for internships, research, or practicum abroad and submit a reflective summary. International students enrolled at KU are eligible and are considered studying abroad at KU. The academic component of the program includes foreign language or courses with an international focus. Students who obtain a grade of C or better in two semesters of any one modern foreign language or international students who pass the English proficiency requirements can document this academic accomplishment. The second academic option is course work with an international focus. Most students take courses in three different departments. Students in professional schools may take two courses in the same department, and international students must take a U.S.-related course. The third component of the program includes co-curricular activities within one of the following categories: event, club, organization, volunteer activity, modern language activity, or other international experiences outside of the classroom.

KU students use an online portfolio to document their progress toward completion of courses and their international or co-curricular experiences. In addition to the completion of the international experience or the co-curricular activity components, students submit a reflective summary of the experience and how study abroad benefited them.

Unpacking Study Abroad

Central Connecticut State University (CCSU) was selected as an exemplar because it has an international focus, and there is a campus-wide continual renewal process to gauge progress toward improving internationalization across the institution. The CCSU mission, values, and strategic plan incorporate international education and the development of culturally and globally aware students. As part of the commitment to internationalization, CCSU participated in the Internationalization Laboratory sponsored by the American Council on Education (ACE), where the campus underwent an intensive review for internationalization.

During a 2012 Association for International Education Administrators (AIEA) conference presentation, Provost Carl Lovitt shared the next steps for internationalization of CCSU via a task force review: faculty-led study abroad, the integration of international education into the curriculum and community engagement, the use of external funds for international efforts, and the infusion of technology for internationalizing the university. The Internationalization Laboratory and the provost's goals guide the internationalization of an institution.

One example of the internationalization of the undergraduate experience is the Unpacking Study Abroad Workshop that students complete after their involvement in a course abroad or semester abroad. The workshop is the culminating activity in which students develop professional applications for what was learned during the overseas experience. Students respond to reflective questions on their attitudes, skills, and knowledge and have the opportunity to explain global events and issues, demonstrate behaviors that model global citizenship, and exhibit respect for intercultural and global diversity. The focus is on what was obtained from the experience that was unique or could not have been achieved otherwise. During mock individual or group interviews, students answer questions about how the study abroad program makes them good candidates for a job and specific cross-cultural skills that were obtained and can be transferred to the workplace.

The workshop provides students with the opportunity to articulate their overseas experiences. Students who study abroad can share how they interact with people from diverse perspectives, adapt to change, and understand cultural differences.

Institutional commitments to internationalization of the campus and undergraduate experiences are both reasons that CCSU was selected. The leadership is dedicated to continued investment in the development of globally and culturally aware students. When students complete study abroad, they reflect on their international experiences and make connections to their professional aspirations. CCSU is one example of this approach to institutional internationalization.

The Global Understanding Course

Offering a campus-wide freshman-level course is another approach to internationalization. The Global Understanding course at East Carolina University (ECU) in Greenville, North Carolina, uses a live video conferencing and e-mail chat to connect students with forty-three partners in twenty-five countries. The program reaches out to partners in countries such as Namibia, Moldova, China, Malaysia, Peru, and Russia.

ECU began offering the course in 2004 and Rosina C. Chia, psychology professor, and Elmer Poe, technology systems professor, spearhead the course. ECU and the partner programs have offered over three hundred courses and enrolled more than five thousand students. The link for this cost-effective, technology-based globalization course is http://www.ecu.edu/cs-acad/globalinitiatives/course.cfm (Chia & Poe, 2012).

During the Global Understanding course, students engage in conversations and complete assignments with three foreign partner programs. The four institutions are paired and switch partners every five weeks so that all students can learn from different cultures during the semester. There are opportunities for students to work with international peers to discuss topics such as family structure and college life via e-mail and for small groups of students to participate in a similar discussion using video conferencing. Web-based reading materials, such as the daily news for each country, guide the discussion. In addition, student partners collaborate through e-mail to write a paper that is submitted to their professors. Through these interactions, students are exposed to diverse perspectives, social norms, cultural traditions, stereotypes, and prejudices.

There are several unique advantages to the course: 1) The course is for freshman-level students so that there is an early opportunity for students to obtain global understandings and potentially engage in future academically focused global experiences. 2) The use of video links is less expensive than other technologies and is accessible in many countries. 3) Students have an opportunity to work in teams with diverse individuals. 4) The success and sustainability of the course are based on a strong leadership team.

The Institute for Global Studies

Undergraduate students use social media and Web-enhanced laptops or smartphones to share their study abroad experience. They post updates to social networks such as travel blogs and Facebook. These networks become a diary of images and information about their social and cultural experiences.

Institutions are beginning to use networking tools for study abroad students to share experiences. For example, the Institute for Global Studies at the University of Delaware (UD) hosts study abroad blogs (http://www.udel.edu/global/studyabroad/). From the Institute for Global Studies blog, which includes diverse summer abroad locations, to specific study abroad programs in countries such as Panama, Vietnam, Brazil, and Tanzania, UD students share insights, observations, and impressions about world cultures. On the Panama 2012 blog, students posted videos about a topic with the audio in Spanish. During the faculty-led trip to Tanzania, students became photojournalists during a guided safari and published Tanzania 2012 UD Study Abroad. On these and other blogs, bloggers document their study abroad experiences, reflect on lessons learned, and develop skills to disseminate information electronically.

While it is challenging to do so, institutional efforts to incorporate social networking into academics are occurring. Educators can guide students as they become professional producers and consumers of technology. More important, the knowledge, skills, and attitudes that students document during study abroad can be used as a reflection on an experience.

Campus Internationalization

Campuses are infusing technology into the international experiences of students. From campus-wide global awareness requirements at KU to study abroad blogs at the UD, institutions have on-campus and off-campus global or international educational experiences for undergraduates. Innovative educators such as the lead faculty at ECU electronically link students around the world. The Global Awareness Program at KU is available to all undergraduates and has an online structure to document the course requirements and international and co-curricular experiences. When CCSU students complete study abroad, they participate in mock interviews to reflect on their international experiences and begin to articulate their knowledge and skills that apply to their professional aspirations. These are specific examples of how institutions are using technology for internationalization. Higher education institutions must continue to access the strengths and needs for the internationalization of the undergraduate experience and develop action plans for its implementation.

Undergraduate Experiences

Undergraduates take global or international courses, participate in study abroad and foreign language programs, or engage in cultural experiences on campus, and it is important for students to discuss their experiences in academic settings. In the KU Global Awareness Program, students write reflections on international and co-curricular activities. During the study abroad programs at the UD, students share their experiences through blogs. At ECU, students work collaboratively across international institutions and learn to communicate with diverse populations. It may not be obvious how the knowledge and skills obtained during international experiences transfer to new experiences or professional settings, but the mock interviews at CCSU do include questions about how study abroad enhanced their skills. At all of the institutions described above, undergraduate students interact with diverse populations and reflect on students' experiences. In addition, institutions provide technical support to electronically document the international experiences of students. With the increase in technology use on campuses, students capture reflections on international experiences using tools such as video conferencing, e-mail, blogs, and e-portfolios.

Ultimately, employers want candidates who can articulate their technology and international knowledge and skills. Below are guiding questions that build on the CCSU mock interview and go beyond describing the international or technological experience or skills of candidates:

1. Why was the specific global experience selected?
2. What was the biggest surprise during the experience?
3. What did the student learn about him or herself?

4. What digital media were used? What was not used? Why?
5. What global or technical experience continues to be a challenge? Why?
6. How do digital media skills enhance his or her ability to communicate and work with others locally or abroad?
7. How has technology and/or an international experience helped the student adapt to change?
8. How has the student adapted to interact with people who are different from him or herself?
9. What digital media and international knowledge or skills does the student have that make him or her qualified for the profession?
10. How will the student maximize the new knowledge and skills in the workplace?

Conclusion

The internationalization of the undergraduate experience is present in higher education, and the use of technology to internationalize the curricula is evident in the examples provided. While these institutions incorporate a range of technologies into internationalization, more can be done to incorporate networking or mobile devices into education. Institutions need strategic plans to implement and assess the infusion of technology into the internationalization of undergraduate education. A review of on-campus, online, or overseas delivery models for international education is needed to inform higher education. Digital content or images from networking sites, e-portfolios, or other data sources from undergraduate global experiences can provide institutions with information on what students upload and their comments and reflections. Evaluation of learning outcomes related to international and technological experiences is needed to document the quality of student work and successful experiences. The dissemination of best practices for using technology to maximize the international experience can provide institutions with teaching and learning methods. A strategic plan to incorporate technology into the internationalizing of the academic experience of undergraduates can transform the education of students and be a catalyst for change in higher education.

Beyond the undergraduate academic experiences at institutions, students engage in extracurricular activities that are international or global. They participate in nonprofit organizations, service learning, or mission trips. Faculty or international experts deliver lectures on global issues such as climate change, cybercrime, and world health. On Facebook and other social networks, students communicate with friends around the world about sports, politics, and other issues of the day. They are developing a lifelong Web presence that can serve to connect higher education and their professional life.

References

Arambewela, R. (2010). Student experience in the globalized higher education market: Challenges and research imperatives. In F. Maringe & N. Foskett (Eds.), *Globalization and internationalization in higher education: Theoretical, strategic and management perspectives* (pp. 155–174). London: Continuum.

Chia, R. C., & Poe, E. (2012). *The global understanding course*. East Carolina University, Greenville, NC. Retrieved from http://www.ecu.edu/cs-acad/globalinitiatives/course.cfm

Dahlstrom, E., de Boor, T., Grunwald, P., & Vockley, M. (2011, October). *The ECAR national study of undergraduate students and information technology [Research report]*. Boulder, CO: EDUCAUSE Center for Applied Research. Retrieved from http:// www.educause.edu/ ecar

Foskett, N., & Maringe, F. (2010). The internationalization of higher education: A prospective view. In F. Maringe & N. Foskett (Eds.), *Globalization and internationalization in higher education: Theoretical, strategic and management perspectives* (pp. 305–318). London: Continuum.

Global awareness program (GAP). *KU International Programs, the University of Kansas*. Retrieved from http://www.international.ku.edu/gap/

Green, M. F., Luu, D., & Burris, B. (2008). *Mapping internationalization on U.S. campuses: 2008 edition*. Washington, DC: American Council on Education.

Institute of International Education. (2011). *Open doors 2011: Study abroad by U.S. students rose in 2009/10 with more students going to less traditional destinations* [Press release]. Retrieved from http://www.iie.org/en/Who-We-Are/News-and-Events/Press-Center/Press-Releases/2011/2011-11-14-Open-Doors-Study-Abroad

Knight, J. (1993). Internationalization: Management strategies and issues. *International Education Magazine, 9*(1), 21–22.

NAFSA Task Force on Internationalization. (2008). *NAFSA's contribution to internationalization of higher education*. Washington, DC: NAFSA, Association of International Educators.

Pew Center's Internet & American Life Project. (2011). *Who's online: Demographics of Internet users*. Retrieved from http:// www.pewinternet.org/Static-Pages/ Trend-Data/Whos-Online.aspx

Qiang, Z. (2003). Internationalization of higher education: Toward a conceptual framework. *Policy Futures in Education, 1*(2), 248–270.

The state of mobile apps. (2010, June 1). *Nielsenwire* [Web log]. Retrieved from http://blog.nielsen.com/nielsenwire/online_mobile/the-state-of-mobile-apps

Study abroad. *Institute for Global Studies, University of Delaware*. Retrieved from http://www.udel.edu/global/studyabroad/

CHAPTER NINE

Global Citizenship and the Complexities of Genocide Education

Antonio J. Castro
Rebecca C. Aguayo

Preparing students as global citizens demands that teachers and students engage in conversations about genocide education, human rights abuses, and ways to dismantle intolerance. Much of genocide education focuses solely on the Holocaust and leaves silent how genocides still occur in today's world. This chapter presents strategies for teaching more recent or current genocides as a way to foster critically minded, global citizens.

"Never again!" "Those who do not remember the past are condemned to repeat it." These catchphrases echo through the halls of American public schools as students learn about the massacre of millions of Jewish people known as the Holocaust (Totten, 2004). Holocaust education, which for many U.S. students represents the only genocide education they receive, impresses on them the gravity of inhumanity and cruelty to others with the intention of conveying a moral imperative to fight against such atrocities. Despite these noble aims, the prevalence of Holocaust education may pose negative consequences for understanding human rights abuses and genocides today. First, Totten (2001) suggested that "if students do not learn about other genocides, they may assume that the Holocaust was simply an aberration of history" (p. 310), something relegated to the historical past. Second, Holocaust education, which refers to the ways in which the Holocaust is presented in classrooms, often promotes narratives that distract from key issues of human rights and social justice. Drawing on the need for a global citizenry, this chapter exposes the shortcomings in U.S. narratives associated with Holocaust education and offers strategies for teaching about modern genocides as a way to foster critically minded global citizens.

Collective Memory and Holocaust Education

When depicting human rights abuses and/or wars, nations often present stories aimed at securing a collective memory about the past by manipulating what is "remembered" and "forgotten" about an event. Collective memories result in "alternative perspectives [being] often lost or suppressed so that the past can be seen in ways that are compatible with dominant values and ideologies" (Leahey, 2010, p. 12). Indeed, within the public schools Crawford and Foster (2007) contended that "official memory becomes dominated by the state to serve cultural and political ends" (p. 6). Textbooks and school curricula often communicate an idealistic or heroic image of the nation (Loewen, 2007). These messages distort historical interpretations that may lead to deeper critical awareness in place of unrealistic and inaccurate accounts of the past.

In public schools, learning about the Holocaust gets subsumed within a larger collective memory of World War II. According to Nicholls (2006), "In U.S. textbooks World War II is depicted as a clean-cut affair, an epic tale of good versus evil with its own heroes and villains" (p. 97). In this collective memory, the United States is cast as a moral hero and Hitler as evil. Totten (2000) discovered the most common misconception about the Holocaust occurred when students attributed its sole cause to Hitler's evil policies. Placed within this collective memory, Holocaust education serves as fodder for establishing the inherent evil of Hitler (and by comparison the moral justification of the actions of the United States and its allies).

This good-versus-evil myth embedded in many textbooks poorly prepares students for global citizenship. First, issues of human rights and genocide become lost in the Holocaust tale, with little investigation of human rights and genocide still occurring today (Totten, 2001). Second, by attributing genocide to the work of only one person or to divine purpose, little if any attention is paid to how genocides arise out of human rights abuses, how they are resisted, and how they are prevented (Shiman & Fernekes, 1999). Third, ways in which Western nations (such as the United States) become either complicit or active supporters of human rights abuses and genocides remain hidden and unexplored (Power, 2002). Finally, not addressing these key historical facts thwarts students' understandings of the complexities of human rights abuses and genocides occurring in the world.

Global Citizenship and Human Rights Education

As societies and schools become more global, education must foster a critical global citizenship for our students. The transition from national to global citizenship, according to Banks (2004), occurs naturally as individuals adopt multiple, complex, interactive, and contextual identities. As a result, individuals can share national and global concerns simultaneously. Thus, "global citizenhip recognizes the fluid, interdependent, and transnational nature of our modern, technologically connected world" (Scarlett, 2009, p. 173). In this way, citizens trace the lines of influence and counterinfluence from local to global contexts and vice versa, seeing how "our actions and behaviors, and the decisions of the government we elect will have an impact, not only on our own lives, but also theirs [global others]" (Osler, 2008, p. 457). Global citizenship implies a deeper sense of awareness or global-mindedness.

Global citizenship also suggests an interrogation of the ways in which individuals, nations, and other entities function within the global community to promote or hinder greater justice and democracy. With regard to human rights education, global citizenship education invites students to investigate the depth and complexity of global injustice and genocide. As Rittner (2004) stated, "We must show students 'the darkest part of humanity,' but also should empower them to 'strive to be the light that drives back the darkness'" (p. 5).

Merryfield (2001) outlined three processes that must occur in social studies education to ready youth for global citizenship. First, students must analyze how the legacy of imperialism affects what is considered legitimate knowledge in today's classrooms. In this process, students and teachers unpack the hidden meanings within textbook narratives, the collective memories and their distortions which are taught about the global other. Having uncovered and critiqued this knowledge base, students and teachers must then develop an understanding of the worldviews of underrepresented peoples. Such knowledge oftentimes remains silenced, oppressed, or downplayed in mainstream texts. Finally, Merryfield called on teachers and students to participate in sustained and reflective cross-cultural experiences. Drawing on models for intercultural awareness, teachers and students can engage in dialogue about lived experiences associated with one's cultural biographies, study the cultural histories of others, and engage in meaningful interactions with culturally different others. In the process of increasing one's cultural awareness, teachers and students also deepen their capacity for global perspective taking.

These processes can lay important foundations for human rights and genocide education, which at its core "teaches the common humanness of the other [that] prepares individuals for doing good acts" (Rittner, 2004, p. 3). Applying notions of global citizenship, we have identified three processes for investigating modern-day genocides that foster this core mission. First, teachers and students must explore how modern genocides result from racism and intolerance. Second, the study of genocide ought to relate how individuals resisted the frenzy of genocidal acts in their community. Finally, the study of modern genocides can reveal how Western nations and their citizens have been complicit or even supported human rights abuses. These investigations foster awareness of the complexities associated with genocide, which is necessary for global citizens to consciously act to eliminate human rights abuses. Here we illustrate each of these three processes as they apply to modern-day genocides.

Learning about Intolerance: The Case of Darfur

In times of limited resources, political disenfranchisement, and economic depression, persistent racial, cultural, and religious intolerance conflicts can ignite into full-blown genocide. Studying genocides must uncover these racial, cultural, and religious tensions and how stereotypes and prejudices both within and outside a country allow opportunities for human rights abuses. Because the atrocities in Darfur are extremely complicated and still recent, we will focus only on a few key elements.

Tensions in Darfur, a part of Sudan, separated two rival groups for centuries. These groups consisted of the "African" or "black" tribes, the Fur, Massalit, Zaghawa, Daju, and Berti, who are mostly sedentary farmers, and the more nomadic Arab herders, who encroached on settled lands. After years of economic and ethnic tensions, General Omar Hassan Bashir led a coup in 1989, marking a turn towards Arab fundamentalism. In response to the coup and feelings of being ignored by the government, "African" Darfuris took up arms against the Islamic government. In retaliation for the armed raid, the Sudanese government trained and employed an Arab militia, the Janjaweed, leading to mass destruction in Darfur. The Janjaweed, aided by the government, bombed, raided, and burned villages; raped women; killed men; abducted children; and destroyed infrastructure to systematically kill or displace the members of the Fur, Massalit, and Zaghawa (Collins, 2006). The atrocities in Darfur have resulted in the death of over two million people and the displacement of millions more (Natsios, 2006).

Stereotypes and prejudices form the root of human rights abuses. Two central ideas arise from the study of Darfur as a case study of intolerance. First, racial intolerance can serve as justification for the dehumanization of the victims of genocide. However, although the ruling governmental parties spread an ideology of intolerance and hatred, these tensions already existed in the background as essential ingredients for human rights abuses. For Darfur, what separated these two groups (African vs. Arab) centered primarily on cultural differences rather than differences in language or religion. Most Darfuris are Muslim and many speak Arabic, like the culturally nomadic Arabs; however, the "African" or "black" tribes of Darfur, the Fur, Massalit, Zaghawa, Daju, and Berti, for example, were likely to be sedentary farmers, rather than nomadic cattle or camel herders. A movement in the 1980s to "Arabize" Sudan and other sub-Saharan areas with the intent of securing its resources exacerbated these cultural differences (Stanton, 2006). In 1987, a group called the "Arab Gathering" wrote to the Islamist government in Khartoum for ideological support, claiming "historical longevity, authenticity, and superiority" of the Arabs in this area of Darfur (Kiernan, 2007, p. 594). Ultimately, these racist beliefs led to the use of the "pejorative epithet *abid*, slave, to distinguish between Arab and African" (Collins, 2006, p. 9) and to the government's ability to justify the killing, rape, and enslavement of these disenfranchised groups. In learning about the powerful role of intolerance, teachers can empower students to act against intolerance of any kind, especially in everyday circumstances, as a way to prevent human rights abuses.

Second, the Darfur case demonstrates how economic disparity magnifies intolerance. Traditionally, the "African" or "black" populations owned land, whereas the nomadic Arabs were granted access to land and water as they moved through certain regions. Increased population growth, droughts, and the rise of poverty disrupted the delicate balance between the farmers and the nomadic herders, leading to conflict over precious resources and polarized one cultural group against the other (Kiernan, 2007). Global citizens who advocate for peace in the world should uncover how economic injustice occurring globally affects the realities of those in countries facing shortages in resources. Like persistent intolerance, economic decline and poverty offer an additional ingredient in human rights abuses and genocides. Tracing these lines of influence from the global to the local with respect to Darfur or other genocides, students can increase civic-mindedness and conscious perspectives surrounding these issues.

Teaching about intolerance can reveal complexities about how human rights and genocides occur within a country and how external global pressures can exacerbate such tensions.

Resisting Genocide: Acts of Courage in Rwanda

Shiman and Fernekes (1999) asserted that students "must understand the dynamics of participation in acts of genocide and other human rights violations if we are to empower youth for human, active, and morally engaged participation in a democracy" (p. 61). Students may often see genocides as senseless acts. In these stories of genocides, Jefremovas (1995) warned, "lie the forgotten stories of humanity: those who struggled and resisted, as well as those who succumbed or fled" (p. 28). An appreciation of these human responses can demonstrate how complex genocides can be.

In the Rwandan genocide, fundamentalist Hutu military, political leaders, and others rallied Hutu civilians in the systematic extermination of Tutsi families. In the frenzy, many Hutus were forced to join in the killing for fear of being killed themselves. The risks of resisting paralyzed many Hutus from taking a stand against the killings. However, acts of resistance did occur. Studying both the consequences of deviances as well as the courageous acts of resistance can help students move beyond dismissing a genocide as senseless and, thereby, irrelevant to students' everyday lives.

First, students ought to explore the complexities of why some resisted the killings of Tutsis and others did not. Uvin (1998) details a variety of explanations for why ordinary people engaged in the killings of the Tutsi. These reasons include frustration as a result of economic recession, political manipulation, opportunity for personal gain, and fear. Exploring these tensions can lead to significant discussion about the complexity of human rights activism and the deeper contexts of the genocide.

Second, studying acts of resistance can dispel simple images of victims as passive and aggressors as senseless. Within the study of resistance, students can uncover themes of the "common humanness" (Rittner, 2004, p. 3) of peoples despite the surrounding atrocities. While the story of Paul Rusesabagina, who managed a hotel during the genocide, is portrayed in the movie *Hotel Rwanda*, a series of smaller acts of resistance reveals the ways in which people tried to resist the genocide. For example, when Hutus invaded a church to kill Tutsis, one Sister called out the leader of the group, who was a member of the church. Her plea convinced the leader to leave the church (Mukarwego, 2010). In another example, a Hutu man charged with killing his Tutsi neighbor threw her onto a pile of dead bodies unharmed, pretending that she was dead. The woman later escaped (Jefremovas, 1995). Countless other stories of resistance abound in the study of the Rwandan genocide.

The goal here is to challenge students to be reflective about the risks and the moral obligations of resistance. Before becoming global citizens, students need to see the common humanity of others even in times of crisis.

Critiquing Western Influences on Genocide: Cambodia as an Outcome of International Politics

Western nations, such as the United States, often hold powerful influence in the persistence of ethnic violence, either directly or indirectly (Power, 2002). Studying the politics of Western involvement in genocides disrupts the collective memory found in school textbooks and curricula. This knowledge reveals how oftentimes citizens are unaware of the political motivations and action of their political leaders within the global arena.

In Cambodia, Pol Pot's rise to power mandated the evacuation of cities and the elimination of political enemies. In order to control the population, he and his Khmer Rouge forbade citizens from feeding themselves through means other than the food rations issued by the state, speaking a foreign language, reminiscing about past times, flirting, praying, owning private property, and making contact with the outside world. Those disobeying were severely beaten or killed. The Khmer Rouge committed genocide aimed at destroying minority populations, Vietnamese, Chinese, Buddhist monks, and Cham Muslims, deposing of almost two million people (Power, 2002).

Geo-political interests and Cold War politics dictated U.S. policy to the Khmer Rouge rather than humanitarian concerns. Before Pol Pot's reign, President Nixon, eager to prevent the spread of Vietnamese communism, extended the Vietnam War into Cambodia. This action had the opposite effect and pushed the Vietnamese farther into Cambodia, which further exacerbated ethnic tensions that had existed since the time of French imperialism. This Vietnamese influx and also the U.S. support for the failed coup of Lon Nol against the established monarchy legitimized the rule of the Khmer Rouge, setting the stage for Pol Pot's domination of Cambodia and the ensuing genocide (Etcheson, 2004; Power, 2002).

Despite hearing news of genocidal acts, U.S. leaders preferred the path of nonintervention because they saw no geo-political advantage to being involved. Later, the United States denounced the Vietnamese invasion of Cambodia, which in effect ended the genocide, and supported Pol Pot's regime militarily and medically despite knowledge of the atrocities it had committed; politically, U.S. leaders considered Vietnam's influence as a greater threat than Cambodia's reign of terror and did not want to risk losing the support of China, a key ally of the Khmer Rouge (Power, 2002).

Education for global citizenship ought to help students see how geopolitical aims often dictate Western intervention, rather than humanitarian concerns. Such an understanding reveals the complexity associated with genocide, intervention, and prevention. Preparing youth with awareness of these issues can lead future citizens to formulate action plans that expose the gap between the stated values of a country and its actions with respect to countries where human rights abuses occur.

The Complexities of Genocide Education

These three approaches to the investigation of genocide—studying the nature of intolerance and racism, exploring resistance, and uncovering Western involvement—offer teachers and students important points for reflection. Genocide education is complex, because genocides and human rights abuses are complex and riddled with contradictions, long histories, and ambiguous relationships between local and global actors. Rittner (2004) called for genocide education that stresses "a 'critical imagination'; that is, an emphasis on developing in students the ability to question assumptions, to challenge what is taken for granted, and to approach knowledge and truth as the student of human invention" (p. 3). These skills debunk collective memories about historical and contemporary events. Such skills are essential for fostering the values of global citizenship in today's youth.

References

Banks, J. A. (2004). Democratic citizenship education in multicultural societies. In J. A. Banks (Ed.), *Diversity and citizenship education* (pp. 3-15). San Francisco: Jossey-Bass.

Collins, R. O. (2006). Disaster in Darfur: Historical overview. In S. Totten & E. Markusen (Eds.), *Genocide in Darfur: Investigating the atrocities in the Sudan* (pp. 3-24). New York: Routledge.

Crawford, K. A., & Foster, S. J. (2007). *War, nation, memory: International perspectives on World War II in school history textbooks.* Charlotte, NC: Information Age.

Etcheson, C. (2004). Case study 7: The Cambodian genocide. In S. Totten (Ed.), *Teaching about genocide: Issues, approaches, and resources* (pp. 169-179). Greenwich, CT: Information Age.

Jefremovas, V. (1995). Acts of human kindness: Tutsi, Hutu and the genocide. *Issue: A Journal of Opinion,* 23(2), 28-31.

Kiernan, B. (2007). *Blood and soil: A world history of genocide and extermination from Sparta to Darfur.* New Haven, CT: Yale University Press.

Leahey, C. R. (2010). *Whitewashing war: Historical myth, corporate textbooks, and possibilities for democratic education.* New York: Teachers College Press.

Loewen, J. W. (2007). *Lies my teacher told me: Everything your American history textbook got wrong.* New York: Touchstone.

Merryfield, M. M. (2001). Moving the center of global education: From imperial world views that divide the world to double consciousness, contrapuntal pedagogy, hybridity, and cross-cultural competence. In W. B. Stanley (Ed.), *Critical issues in social studies research for the 21st century* (pp. 179-207). Greenwich, CT: Information Age.

Mukarwego, M. C. (2010). Some church leaders did what they could to help the innocent. In A. Cruden (Ed.), *The Rwandan genocide* (pp. 130-138). Detroit, MI: Gale.

Natsios, A. S. (2006). Moving beyond the sense of alarm. In S. Totten & E. Markusen (Eds.), *Genocide in Darfur: Investigating the atrocities in the Sudan* (pp. 25-42). New York: Routledge.

Global Citizenship and the Complexities of Genocide Education

Nicholls, J. (2006). Beyond the national and the transnational. In S. J. Foster & K. A. Crawford (Eds.), *What shall we tell the children? International perspective on school history textbooks* (pp. 89-112). Greenwich, CT: Information Age.

Osler, A. (2008). Human rights education: The foundation of education for democratic citizenship in our global age. In J. Arthur, I. Davies, & C. Hahn (Eds.), *The Sage handbook of education for citizenship and democracy* (pp. 455-467). Los Angeles: Sage.

Power, S. (2002). *"A problem from Hell": America and the age of genocide.* New York: Harper P erennial.

Rittner, C. (2004). Education about genocide, Yes: But what kind of education? In S. Totten (Ed.), *Teaching about genocide: Issues, approaches, and resources* (pp. 1-5). Greenwich, CT: Information Age.

Scarlett, M. H. (2009). Imaging a world beyond genocide: Teaching about transitional justice. *The Social Studies, 100*(4), 169-176.

Shiman, D. A., & Fernekes, W. R. (1999). The Holocaust, human rights, and democratic citizenship education. *The Social Studies, 90*(2), 53-62.

Stanton, G. H. (2006). Proving genocide in Darfur: The atrocities documentation project and resistance to its findings. In S. Totten & E. Markusen (Eds.), *Genocide in Darfur: Investigating the atrocities in the Sudan* (pp. 181-188). New York: Routledge.

Totten, S. (2000). Student misconceptions about the genesis of the Holocaust. *Canadian Social Studies, 34*(4), 81-84.

Totten, S. (2001). The "null curriculum": Teaching about genocides other than the Holocaust. *Social Education, 65*(5), 309-313.

Totten, S. (2004). Introduction. In S. Totten (Ed.), *Teaching about genocide: Issues, approaches, and resources* (pp. vii-xi). Greenwich, CT: Information Age.

Uvin, P. (1998). *Aiding violence: The development of enterprise in Rwanda.* Hartford, CT: Kumarian Press.

Part 2

Global Issues and Innovative Instructional Practice for Teaching Global Education

CHAPTER TEN

Broadening Horizons: Utilizing Film to Promote Global Citizens of Character

Stewart Waters
William B. Russell III

> Movies can and do have tremendous influence in shaping young lives in the realm of entertainment towards the ideals and objectives of normal adulthood.
> Walt Disney (1901-1966)

In March 2012, the Invisible Children Organization released a thirty-minute video to raise awareness about Joseph Kony, the infamous leader of the Lord's Resistance Army (LRA) in Uganda. In a matter of weeks, this well-designed film went viral, generating over one hundred million views. Common citizens, celebrities, professional athletes, and politicians all began sharing their support for this movement and for the objective of capturing Joseph Kony. Social media sites such as Facebook and Twitter became political forums for people to share this video and their thoughts on an international issue of human rights. However, the immense popularity of this film brings to light several important questions. What made this film so intriguing to people? How much did people already know about Joseph Kony prior to watching the film? How do people from Uganda feel about this film? These and many other questions stand to initiate some fairly interesting, and certainly controversial, discussions about global human rights and the responsibility people have to look after the well-being of others. Since the film's purpose was to raise awareness of Joseph Kony and also encourage people everywhere to take action supporting the objective of capturing Kony by the end of 2012, it will be interesting to see how active viewers of this film actually are in this pursuit. There is already some evidence to support the notion that young people genuinely seem concerned about this issue. Take, for example, a recent Skype interview that Senator Johnny Isakson, R-Ga., had with a sixth-grade class from Westside Middle School in Winder, Georgia. The very first question these students posed to the senator was, "Are you doing anything about Joseph Kony?" (Cassata, 2012). Dustin Davis, the students' teacher, commented on how the students viewed the video on their own and brought it up as a topic of discussion in class, some being reduced to tears. Mr. Davis realized the importance of this topic to the students and incorporated it into his curriculum as part of the goal to "teach them the role of a citizen in our country" (Cassata, 2012). Regardless of how people feel about this political video or the Invisible Children Organization that sponsored it, one thing is for sure: the "Kony 2012" film clearly demonstrates the power of cinema to generate meaningful and reflective discussions about making character-based choices in a global society.

The purpose of this chapter is to examine how films can be used to facilitate relevant and powerful discussions in the classroom to encourage the development of global citizens of character. The National Council for the Social Studies believes the goal of global citizenship should be "to develop in youth the knowledge, skills, and attitudes needed to live effectively in a world possessing limited natural resources and characterized by ethnic diversity, cultural pluralism, and increasing interdependence" (National Council for the Social Studies, 2005). As the aforementioned example indicates, students today are very interested in global issues, particularly when these issues are presented through the dynamic and emotional realm of film. Global issues of human rights, injustice, war, and suffering are no longer words in textbooks for students. Film provides the opportunity to visualize and relate to the emotional plights of others in a very personal way. There is little debate in the twenty-first century regarding the interconnectedness and interdependence of people all over the world. Increased globalization has brought with it a myriad of changes in social, political, and economic circles. But these changes are not always readily understood or recognized by many students who may struggle to see the connections between their lives and people in other countries.

Educators must continually work to help students understand that they are viewing the world and society through a specific lens, influenced by several social, cultural, and political factors. In addition, students should be provided with opportunities to discuss the lenses of "others" that may come from very different situations and how these people may interpret the same events based on their perspective. Films are the ideal resource for this type of analysis, since people often have differences of opinions regarding the same film. While the goal of helping students to see things from multiple perspectives to improve global citizenship concepts is undoubtedly valuable, teachers may struggle to realize how and where these lessons can fit into an already crowded social studies curriculum. However, such lessons foster a global citizenship based on strong character values, which include respect, responsibility, and understanding, to name just a few.

Character education programs and initiatives offer a strong curricular framework for the inclusion of global citizenship initiatives. Schools all over the world, not just in the United States, are currently grappling with the dilemma of helping students understand their mutual roles as citizens of a country and as citizens in the world community. For this reason, character education and global citizenship share a unique presence as a foundational goal of education all over the world. One example of this worldwide interest in character education is the First Annual International Conference on Character Education, held in 2011 at Yogyakarta State University, Indonesia. Recognizing the importance of character education abroad can help many U.S. teachers to view character education not just as a "special interest" topic in the curriculum, one that will soon disappear like many other educational initiatives. Rather, character education has demonstrated a substantial amount of perseverance over the years and could serve as the curricular catalyst needed to infuse global citizenship goals into the classroom (Lockwood, 2009).

Developing Global Citizens of Character Through Film

In the United States, the film industry produces countless popular films every year depicting a myriad of character issues and a variety of biases when portraying the lives of people from other cultures. By critically examining these films, teachers provide students the opportunity to go beyond the surface level, beyond traditional depictions of other cultures. As Merryfield (2011) points out, "'Traditional' cultural attributes (traditional dance, music, food, clothes, holidays or lifestyles) misleads students as it is not what they would see if they visited the country today" (p. 64). Acknowledging biases in film representation of cultures allows teachers to pose such questions as, "Why do you think this culture was presented in this way?" and "How do you think members of this group would feel about how they are portrayed in the film?" These questions, although tough, offer students a strong and relevant foundation for considering their place in a global world.

One of the pivotal aspects of this approach critical to successful classroom implementation is shifting how students view films. While the primary purpose of films is entertainment, there still remains a great deal that can be learned from films when viewed critically. As Salomon (1994) points out, "the focus has to shift from what students watch to how they watch" (p. 4). As the world continues to become more interconnected via new technology, the importance of being critical consumers of information becomes paramount because there is more information available than ever before. Students do not have the luxury of assuming that what they see or read is reliable or objective. If teachers help students evaluate and interpret meaning from visual materials at an early age, then these students will be prepared to handle the barrage of information provided to them via news reports, films, Webcasts, advertisements, and so forth on important issues that they face as future citizens in a global society.

Another key to developing global citizens of character through films is the decision-making and discussion process. Decision making is a skill that transcends content of school curriculum and finds itself at the core of all education. When showing films, or clips from films, teachers can help provoke meaningful inquiry through thinking about social issues, personal values, and moral dilemmas, thus allowing students to reflect personally and make insightful decisions, which is a key characteristic of being a citizen of character in a global society. Oftentimes, films are very effective in presenting the complexity of character choices and decision making by providing students with situational context and scenarios to see how character-based decisions have influenced historical and contemporary events. Take, for instance, the concepts of responsibility and citizenship as explored in the film *Gandhi* (1982). In this film, Mohandas Gandhi becomes disillusioned with discriminatory and biased laws against Indians living under the control of the British Empire. He initiated nonviolent protests and campaigns to urge the British government to change these discriminatory laws. In spite of facing constant harassment and imprisonment, Gandhi eventually succeeds in helping India win its freedom, only to see violence erupt within his country between Hindus and Muslims.

Questions to consider in this film include the following: What does it mean to be a good citizen? Do good citizens always obey the laws? Can being a responsible citizen sometimes mean breaking the law? What does it mean to be a good citizen in a country with conflicting definitions of this term? It is also worth noting that this film was an international collaboration between production companies in India and the United Kingdom in an effort to be culturally sensitive to both sides during the project. This is just one brief example of how films can introduce and encourage discussion about value conflicts and help students examine the complexity of character-driven decision making in the context of the global world. By critically viewing films in this way, students have the opportunity to reflect on their own values, decision making, and civic responsibilities.

Teaching with film is a powerful instructional method that can increase student interest in the content as well as promote students' use of higher-order thinking skills (Russell, 2012a, 2012c). It is imperative that teachers keep in mind that film is an instructional tool, and like all tools, its effectiveness is determined by its application. When used properly, film can enhance the curriculum by providing students with unique scenarios and context necessary to truly analyze character concepts that are mostly abstract. When used incorrectly, film can be viewed as a way to pass the time, control behavior, or simply entertain the students as a type of reward. Teachers must always remember that the showing of films must be accompanied with learning goals and objectives based on the individual needs of students and any required educational standards. As research has shown, authentic activities, such as teaching with film, help teachers achieve instructional goals such as retention, understanding, reasoning, and critical thinking (Driscoll, 2005).

Based on the vital importance of proper procedures when showing films, the authors offer the Russell Model for Using Film (Russell, 2004, 2007). This four-stage model provides teachers with a general outline of procedures to follow when showing films to make sure teachers effectively and efficiently use films as a learning tool.

Russell Model for Using Film in the Classroom

Stage 1: The Preparation Stage. The preparation stage is the most important stage of the Russell Model for Using Film. This is the planning stage of the model. The preparation stage includes creating lesson plans that incorporate a film, while still meeting instructional goals/objectives, state standards, and national standards, and adhering to all legal requirements. Remember to obtain permission from administration and parents prior to showing a film.

Stage 2: The Pre-viewing Stage. The pre-viewing stage is completed prior to students' viewing the film. The pre-viewing stage should include an introduction of the film and the purpose for viewing the film.

Stage 3: The Watching the Film Stage. The watching the film stage is the period when students actually view the film. The watching the film stage includes watching the film (entirety or clips) and ensuring that students are aware of what they should be doing and looking for while watching the film.

Stage 4: The Culminating Activity Stage. The culminating activity stage is done after students have watched the film. The culminating activity stage includes assessing student learning. After stopping the film, teachers can focus on reviewing, clarifying, and/or discussing major points, concepts, issues, scenes, and/or inaccuracies. Teachers should also assess student learning in some fashion. Possible assessment strategies can include class discussion, class debate, rewriting the ending of the film, writing a review of the film, taking a test/quiz, completing a written assignment aligned with the film and topic/unit, reenactment (have students reenact a scene from the film), having students conduct a mock interview with the star, director, and/or producer of the film, and/or having students analyze and evaluate the film.

Legal Issues

Teachers can use copyrighted films in their classrooms. However, there are legal guidelines to which teachers must adhere when using copyrighted material. Section 110 (1) of Title 17 of the United States Code on Copyright and Conditions (1976) cites the following exemption for the use of copyrighted films for educational purposes:

> Performance or display of a work by instructors or pupils in the course of face-to-face teaching activities of a nonprofit educational institution, in a classroom or similar place devoted to instruction, unless in the case of a motion picture or other audiovisual work, the performance, or the display of individual images, is given by means of a copy that was not lawfully made under this title, and that the person responsible for the performance knew or had reason to believe was not lawfully made.

In short, teachers must adhere to the following guidelines when using films:

- Films must be shown in a nonprofit educational institution. Within the institution, the film must be shown in a classroom or place intended for instruction.
- Films must be for planned educational purposes, not for extracurricular entertainment.
- The teacher must show films to the students in a "face-to-face" encounter.

In addition, teachers are given less restriction for using copyrighted materials in the classroom as part of the Teach Act, signed by President George W. Bush on November 2, 2002. Teachers are absolutely prohibited from making or showing a pirated copy of a film, using film for public performance, and/or from making a profit from the film (Russell, 2012b).

The Three Rs of Film Selection

The films highlighted in this section were all chosen specifically for global citizenship education based on the following three factors:

1. *Relevance*—any age-appropriate film that addresses the personal, social, and moral dilemmas common of the defined age group.
2. *Relatability*—the personal, social, or psychological connection a student makes with the film based on the storyline, lessons, or characters in the film.

3. *Rating*—film ratings determined by the Motion Picture Association of America. (Russell & Waters, 2010)

The films in the following section are divided into sections based on grade level (elementary, middle, and high school). The films are listed in alphabetical order and each film entry includes a film synopsis, Global Citizens of Character Lesson Topics, and the following bibliographic information: year, genre, Motion Picture Association of America (MPAA) rating, alternative title/s, director, producer, length, language, color, company, and cast. The films were selected based on the Three Rs of film selection mentioned above, and also had international emphasis, instructional potential, availability, and connection to the overall objective for building global citizens of character. It is important to note that the short list of films provided is not meant to indicate that these are the only films that can be used for building global citizens of character. Rather, the filmography is meant to provide educators with an excellent starting point of films that have valuable character lessons for students. Summaries of all films are provided to give educators an idea of what the film is about and what values or character lessons each film addresses. Although each of these films is likely to address many different values and character issues, the authors include Global Citizens of Character Lesson Topics at the bottom of each film summary as an example of character topics addressed by the films. In addition, we have included a section of Globally Guided Questions specific to each film in order to offer educators an example of how to analyze the content of each film for questions that could facilitate meaningful discussions in the classroom. As with any type of classroom instruction, teachers need to ensure that the curriculum and instruction meet the learning goals/objectives and the individual needs of students.

While the films listed below are popular and relevant in the United States for addressing global character issues, teachers in other countries can still make use of this method. If the films provided do not meet the needs of your students or community, we encourage all teachers to consider popular films in their countries and to use these resources to engage students in lively discussions regarding the formation of global character development.

Films for Building Global Citizens of Character in Elementary Classrooms

Aladdin (1992)
   Genre: Animation
   MPAA Rating: G

Directed by: Ron Clements and John Musker
Produced by: Ron Clements and John Musker
Written by: Ron Clements and John Musker
Color: Color
Length: 90 minutes
Language: English
Production Company: Walt Disney Feature Animation
Cast: Scott Weinger, Robin Williams, Linda Larkin, Jonathan Freeman, Frank Welker, Gilbert Gottfried, and Douglas Seale

Summary: Aladdin is an interesting character education film to analyze because it deals directly with issues of consumerism, power, and authority. Aladdin is the main character of the film; he is from the streets and falls in love with the beautiful Princess Jasmine. Unfortunately, the princess can only marry a prince, so Aladdin must find a way to be with the woman he loves. In his quest to be with Jasmine, Aladdin becomes entangled with the evil Jafar and a magical genie that has the power to make wishes come true, but at what cost? While this film is both funny and entertaining, it is also filled with moral dilemmas and character lessons that are perfect to help students analyze their own values.

Global Citizens of Character Lesson Topics:
citizenship, fairness, respect, honesty, and caring

Globally Guided Questions:
1. Who makes the rules in society and should you always obey the rules?
2. What is the difference between a need and a want?

An American Tail (1986)
   Genre: Animation
   MPAA Rating: G
   Directed by: Don Bluth
   Produced by: Don Bluth, Gary Goldman, and John Pomeroy
   Written by: Judy Freudberg
   Color: Color
   Length: 80 minutes
   Language: English
   Production Company: Amblin Entertainment
   Cast: Erica Yohn, Nehemiah Persoff, Amy Green, Phillip Glasser, Christopher Plummer, John Finnegan, Dom DeLuise, and Neil Ross

Summary: This Oscar-nominated animated film follows the journey of a Russian family of mice, immigrating to the United States in an attempt to escape the abusive cats of their home country. During their journey, the family loses their son, Fievel. Fievel spends the remainder of the film making friends and trying to find his family, while also avoiding the cats he thought would not be in America. His adventures are meaningful for analysis because they explore such character dilemmas as whom to trust and how to respond in the face of adversity. It is also important to note that this film's popularity prompted a sequel, *An American Tail: Fievel Goes West*, which is also a quality film to use for the purposes of values analysis in character education.

Global Citizens of Character Lesson Topics:
responsibility, caring, perseverance, fairness, and integrity

Globally Guided Questions:
1. What does it mean to be an immigrant and how might life be different for these people?
2. What does it mean to be trustworthy and how does one decide whom to trust?

Fiddler on the Roof (1971)
   Genre: Drama
   MPAA Rating: G
   Directed by: Norman Jewison
   Produced by: Norman Jewison
   Written by: Joseph Stein
   Color: Color
   Length: 181 minutes
   Language: English
   Production Company: Cartier Productions
   Cast: Topol, Norma Crane, Leonard Frey, Molly Picon, Paul Mann, Rosalind Harris, Michele Marsh, Neva Small, and Candy Bonstein

Summary: Fiddler on the Roof is a three-time Oscar-winning film based on the book by Sholom Aleichem. It is the story of a Jewish family living in Russia during the early 1900s. The father, Tevye, is the village milkman who struggles with the responsibility of raising his five daughters and following tradition. Tevye is a hardworking man who wants the best for all of his daughters, but as times change, Tevye is forced to make a series of decisions with complicated outcomes that make him realize the changes needed within himself. This film is excellent for analyzing family relationships, marriage, the responsibilities of a father, and dealing with the inevitability of change.

Global Citizens of Character Lesson Topics:
caring, responsibility, honesty, fairness, and self-discipline

Globally Guided Questions:
1. What does it mean to be Jewish?
2. How do family responsibilities identified in this film relate to those in your community?

Mulan (1998)
    Genre: Animation
    MPAA Rating: G
    Alternative Title: China Doll and the Legend of Mulan
    Directed by: Tony Bancroft and Barry Cook
    Produced by: Pam Coats
    Written by: Rita Hsiao, Chris Sanders, Raymond Singer, and Philip LaZebnik
    Color: Color
    Length: 88 minutes
    Language: English and Mandarin
    Production Company: Walt Disney Feature Animation
    Cast: Miguel Ferrer, Harvey Fierstein, James Hong, Pat Morita, Eddie Murphy, Frank Welker, Donny Osmond, and Freda Foh Shen

Summary: Mulan is an Oscar-nominated film of the retelling of an ancient Chinese folktale about a young woman who fights for her country. As the Huns invade China, one man from every family is summoned to fight for the imperial army. However, Mulan's father has an old wound and is certainly in no condition to fight, but he chooses to join anyway to maintain the honor of his family. Mulan, hoping to keep her father from death in battle, decides to disguise herself as a man and serve in her father's place. With the help of a magical dragon named Mushu, Mulan may have the opportunity to fight for her country and help save China from the Huns. But can she do this while maintaining the secrecy regarding her gender and true identity? This film is open to the analysis of such values as gender roles, love, sacrifice, family, and service to one's country.

Global Citizens of Character Lesson Topics:
patriotism, courage, citizenship, responsibility, and honesty

Globally Guided Questions:
1. Should boys and girls have the same responsibilities?
2. What does it mean to be patriotic and how do people around the world demonstrate patriotism?

The King and I (1956)
    Genre: Musical
    MPAA Rating: G
    Alternative Title: Rodgers and Hammerstein's The King and I
    Directed by: Walter Lang
    Produced by: Charles Brackett
    Written by: Ernest Lehman
    Color: Color
    Length: 133 minutes
    Language: English
    Production Company: Twentieth Century-Fox Film Corporation
    Cast: Deborah Kerr, Yul Brynner, Rita Moreno, Martin Benson, Terry Saunders, Robert Banas, Carlos Rivas, and Rex Thompson

Summary: *The King and I* is a five-time Oscar-winning musical film, based on the book by Margaret Landon, set in nineteenth-century Siam. The story is about Englishwoman Anna Leonowens who, following the death of her husband, travels to Siam in order to teach the king of Siam's children English. Upon arrival in this distant and strange land, Anna finds herself uncomfortable with many of the local customs and traditions but loves the children with whom she works. Her relationship with the king of Siam is also complicated because he views her initially as one of his servants, while Anna considers herself a free woman under the employment of the king. Over time, these two very different characters actually begin to care deeply for each other, with the king wanting Anna to join his harem. Anna's English traditions and values will prevent her from accepting this offer, but her affection for the king and his family may keep her in their lives after this story ends. One of the great character education aspects of this film is the ability to analyze traditional European values in the context of the Asian world.

Global Citizens of Character Lesson Topics:
caring, responsibility, fairness, courage, and integrity

Globally Guided Questions:
1. How do you define traditions and values? Why is it significant to understand the traditions and values of different cultures?

2. When encountering people with different traditions or values, why is it important to respect their culture?

Films for Building Global Citizens of Character in Middle School Classrooms

Cool Runnings (1993)
    Genre: Adventure
    MPAA Rating: PG
    Directed by: Jon Turteltaub
    Produced by: Dawn Steel
    Written by: Lynn Siefert, Tommy Swerdlow, and Michael Goldberg
    Color: Color
    Length: 98 minutes
    Language: English
    Production Company: Walt Disney Pictures
    Cast: John Candy, Malik Yoba, Leon, Doug E. Doug, Rawle D. Lewis, Raymond J. Barry, Peter Outerbridge, Paul Coeur, and Charles Hyatt

Summary: *Cool Runnings* is a fun, spirited comedy based on the true story of the first Jamaican bobsled team. The story begins at the Olympic trials in 1988 when an unfortunate accident keeps a few men from achieving their dreams of competing in the Olympics. One of the men, Derice Bannock, is the son of a former gold medal winner and is determined to compete in the Olympics. While exploring his options, Derice learns that one of his father's old friends is a former bobsled competitor and is on the island. He convinces the coach to build a Jamaican bobsled team to compete in the Winter Olympics. Although the men are initially mocked and ridiculed by the other teams, their toughness and dignity will gain them the respect of people all over the world. This film is great for analyzing character themes such as forgiveness, redemption, and triumph in the face of adversity.

Global Citizens of Character Lesson Topics:
respect, patriotism, self-discipline, perseverance, and courage

Globally Guided Questions:
    1. How does this film depict Jamaican culture?
    2. What stereotypes did the characters in this film face and how did they deal with this adversity?

Gallipoli (1981)
    Genre: Drama
    MPAA Rating: PG
    Directed by: Peter Weir
    Produced by: Patricia Lovell and Robert Stigwood
    Written by: Peter Weir and David Williamson
    Color: Color
    Length: 110 minutes
    Language: English
    Production Company: Australian Film Commission
    Cast: Mel Gibson, Mark Lee, Bill Kerr, Harold Hopkins, Heath Harris, Gerda Nicolson, Ron Graham, and Charles Lathalu Yunipingli.

Summary: This film is about a group of Australian men who join the army to help the British fight during World War I. Two of the men, Archy Hamilton (Mark Lee) and Frank Dunne (Mel Gibson), are well-known track sprinters with bright futures. However, they have a desire to make a difference, so the two men forsake their track futures in order to join the army. The men are trained and eventually sent to fight at Gallipoli, where the tragedies of war and the bonds of friendship will become realities for the Australian volunteers.

Global Citizens of Character Lesson Topics:
patriotism, integrity, courage, responsibility, and perseverance

Globally Guided Questions:
1. What does it mean to sacrifice and how was this demonstrated by characters in the film?
2. Do citizens' responsibilities change during times of war? How and why?

Gandhi (1982)
    Genre: Biography
    MPAA Rating: PG
    Directed by: Richard Attenborough
    Produced by: Richard Attenborough
    Written by: John Briley
    Color: Color/Black & White
    Length: 188 minutes
    Language: English
    Production Company: Columbia Pictures Corporation

Cast: Ben Kingsley, Candice Bergen, Martin Sheen, Edward Fox, Trevor Howard, John Gielgud, John Mills, and Ian Charleston

Summary: This film is a biography of Mohandas Gandhi, one of the most famous human rights activists in history. As a former lawyer, this Indian man introduced the use of nonviolent protests to try and free his country from rule under the British Empire. This film could be used as a case study, looking at how Gandhi exemplified character traits such as responsibility, integrity, respect, courage, and patriotism, and at how these values affected his life and the attitude of other people in his country.

Global Citizens of Character Lesson Topics:
citizenship, fairness, patriotism, courage, and self-discipline

Globally Guided Questions:
1. Can breaking the law be considered patriotic? How? When?
2. How do you define integrity? How do characters in this film display integrity?

The Iron Giant (1999)
    Genre: Animation
    MPAA Rating: PG
    Directed by: Brad Bird
    Produced by: Allison Abbate and Des McAnuff
    Written by: Tim McCanlies
    Color: Color
    Length: 86 minutes
    Language: English
    Production Company: Warner Bros. Animation
    Cast: Jennifer Aniston, Harry Connick Jr., Vin Diesel, James Gammon, Christopher McDonald, John Mahoney, Cloris Leachman, and M. Emmet Walsh

Summary: Set in the United States in the 1950s, *The Iron Giant* is a story about fear and prejudice. A giant alien robot is sent to Earth and is befriended by a young boy named Hogarth Hughes. Hogarth is somewhat of an outcast at school and quickly befriends the Iron Giant. As government agents arrive to find out what exactly came to Earth, Hogarth must try to help his friend. One of the agents, Kent Mansley, is convinced that the Iron Giant is a weapon sent to destroy human life, and he will stop at nothing to kill the Iron Giant. Although this film has many key themes to analyze, it is especially interesting to have students evaluate how fear and paranoia can cloud good judgment and decision making.

Global Citizens of Character Lesson Topics:
affection, control, fairness, friendliness, and generosity

Globally Guided Questions:
1. What does it mean to be prejudiced and how did the film address this concept?
2. Why do people in various cultures throughout human history have a fear of things they do not understand?

The Truman Show (1998)
   Genre: Drama
   MPAA Rating: PG
   Directed by: Peter Weir
   Produced by: Edward S. Feldman, Andrew Niccol, Scott Rudin, and Adam Schroeder
   Written by: Andrew Niccol
   Color: Color
   Length: 103 minutes
   Language: English
   Production Company: Paramount Pictures
   Cast: Jim Carrey, Laura Linney, Noah Emmerich, Natascha McElhone, Brian Delate, Judy Clayton, Ed Harris, Paul Giamatti, and Adam Tomei

Summary: This Oscar-nominated film follows the life of Truman Burbank, an ordinary man who works for an insurance company. Truman is married and has friends throughout his community. What Truman does not know is that his entire life is a reality television show. People around the world watch as Truman lives his daily life, not knowing that everyone around him are actors in this show. When Truman becomes suspicious of his life, he wants to leave the town and venture out into the world. However, the television producers do not want this to ruin their show, and they will stop at nothing to keep Truman in the staged town. Although this film is quite humorous, it does deal with issues such as the effect corporate America may have on our lives, reality television, personal freedom, and the nature of what is considered entertaining to American culture.

Global Citizens of Character Lesson Topics:
freedom, thoroughness, deceit, watchfulness, and control

Globally Guided Questions:
1. How does this film address issues of American consumerism and mass media?
2. What would you consider to be the positive and negative consequences of reality television on American culture? How do these types of shows affect the perception of the United States in different countries?

Films for Building Global Citizens of Character in High School Classrooms

Bend It like Beckham (2002)
    Genre: Comedy
    MPAA Rating: PG-13
    Directed by: Gurinder Chadha
    Produced by: Gurinder Chadha and Deepak Nayar
    Written by: Gurinder Chadha and Guljit Bindra
    Color: Color
    Length: 112 minutes
    Language: English
    Production Company: Kintop Pictures
    Cast: Parminder Nagra, Keira Knightley, Jonathan Rhys Meyers, Anupam Kher, Frank Harper, Juliet Stevenson, and Ameet Chana

Summary: The film is about two very different teenage girls living in London. Jesminder Bhamra and Juliette Paxton both come from very different backgrounds, religions, and families, but they do share a common dream of playing professional women's soccer in the United States. Although the girls are very talented at the game, their parents do not support them and try to encourage a more traditional role for their daughters. Without the support of their parents, the two girls decide to bend the rules in order to realize their dream of playing professional soccer. This film is interesting to use to examine character issues about responsibilities, expectations, and family support for goals.

Global Citizens of Character Lesson Topics:
freedom, happiness, inspiration, obedience, and liberation

Globally Guided Questions:
1. What is meant by gender stereotypes and do you think they exist in all cultures?

2. How do family expectations affect the lives and futures of children? What do you think are the positive and negative influences on family expectations?

Cry, the Beloved Country (1995)
    Genre: Drama
    MPAA Rating: PG-13
    Directed by: Darrell Roodt
    Produced by: Anant Singh and Harry Alan Towers
    Written by: Ronald Harwood
    Color: Color
    Length: 106 minutes
    Language: English
    Production Company: Alpine Pty Limited
    Cast: James Earl Jones, Richard Harris, Charles S. Dutton, Dolly Rathebe, Tsholofelo Wechoemang, and Jennifer Steyn

Summary: This powerful film is a story about two men in South Africa. One is a church minister named Steven Kumalo who goes to Johannesburg to help his son who has been imprisoned in connection with a robbery that ultimately left a white man killed. The father of the white man is a strong supporter of the separation of races in the country at this time. On a journey for the truth, both of these men begin to realize that their sons are complex individuals and gain a deeper insight into their own perceptions of humanity. This is a great film to show students for its topics of race relations, integrity, and understanding.

Global Citizens of Character Lesson Topics:
caring, compassion, devotion, justice, and fairness

Globally Guided Questions:
1. Why do you think race discrimination exists and how can people work to eliminate these problems?
2. How would you define humanity and how is this concept developed and demonstrated throughout this film?

Hotel Rwanda (2004)
    Genre: Drama
    MPAA Rating: PG-13
    Directed by: Terry George
    Produced by: Terry George and A. Kitman Ho
    Written by: Keir Pearson and Terry George

Color: Color
Length: 121 minutes
Language: English
Production Company: United Artists
Cast: Don Cheadle, Nick Nolte, Tony Kgoroge, Rosie Motene, Joaquin Phoenix, Mosa Kaiser, Cara Seymour, and Simo Mogwaza

Summary: This touching and disturbing drama was nominated for three Oscars. It tells the true story of Paul Rusesabagina (Don Cheadle), who is the manager of a hotel in Rwanda when the Hutu militia begins assaulting the Tutsi. The genocide that ensued is one of the worst, most brutal displays of mass execution that the world has ever seen. As the Tutsi people run for their lives, Paul begins providing them refuge at the hotel, putting the lives of himself and his family at great risk. This film is excellent for addressing the complexity of character decisions and how having good personal character is not always easy.

Global Citizens of Character Lesson Topics:
responsibility, citizenship, sympathy, courage, and selflessness

Globally Guided Questions:
1. How do you define "citizenship" and do people around the world have an inherent obligation to intervene during times of war, extreme poverty, natural disasters, or crimes against humanity?
2. How did the character of Paul Rusesabagina influence not only his actions, but also the lives of others?

School Ties (1992)
Genre: Drama
MPAA Rating: PG-13
Directed by: Robert Mandel
Produced by: Stanley R. Jaffe and Sherry Lansing
Written by: Dick Wolf
Color: Color
Length: 106 minutes
Language: English
Production Company: Paramount Pictures
Cast: Brendan Fraser, Matt Damon, Ben Affleck, Chris O'Donnell, Randall Batinkoff, Cole Hauser, Anthony Rapp, and Amy Locane

Summary: This film is about a young man named David Green (Brendan Fraser) who attends an elite boarding school on a football scholarship. Green is from a working-class family and does not relate well to his wealthy classmates. Once one of his teammates finds out that Green is Jewish, prejudice and discrimination ensue. Set in 1950s America, this film explores religious discrimination and how personal character, good and bad, can greatly affect the lives of others.

Global Citizens of Character Lesson Topics:
determination, responsibility, diversity, dignity, and honesty

Globally Guided Questions:
1. Does socio-economic status and social class affect discrimination and prejudices? How? Why?
2. How do you think David Green responded to the discrimination he faced? Would you have handled the situation differently? How? Why?

Traitor (2008)
    Genre: Crime
    MPAA Rating: PG-13
    Directed by: Jeffrey Nachmanoff
    Produced by: Don Cheadle, David Hoberman, Kay Liberman, and Todd Liberman
    Written by: Jeffrey Nachmanoff
    Color: Color
    Length: 114 minutes
    Language: English and Arabic
    Production Company: Overture Films
    Cast: Don Cheadle, Guy Pearce, Said Taghmaoui, Neal McDonough, Alyy Khan, Archie Panjabi, Jeff Daniels, and Lorena Gale.

Summary: Samir Horn (Don Cheadle) is a former member of the U.S. Army Special Forces and a devout Muslim who is working with terrorists throughout the Middle East. Samir has a vast knowledge of explosives and proves to be a valuable asset to the terrorist group. However, Samir is also being followed by FBI agents who want to bring him in for his treasonous acts against the United States. But is Samir really a traitor? This film does a very nice job of presenting the complexities of character issues, citizenship, religious beliefs, and loyalty.

Global Citizens of Character Lesson Topics:
heroism, terrorism, making a difference, sacrifice, and willingness

Globally Guided Questions:
1. How do you think this film deals with the complexities of terrorism? Accurately? Fairly? Explain.
2. How has the increased globalization of society changed what it means to be a good citizen?

Conclusion

Despite all of the different terminologies used for the enterprise of teaching students values—moral education, democratic education, global education, citizenship education, and character education—there is one common goal: training students how to become effective citizens in society. Studying the long and complex history of educating students for citizenship in America reveals the nature of this goal. First, it is undeniable that preparing students for life in society as adults has always been a responsibility bestowed upon teachers and schools. However, as the population, power, and culture of this democratic society have changed over the years, so too has the perception of what it means to be a responsible citizen. Rapoport (2012, p. 81), highlights the shifting perspective of citizenship in the global context, stating that:

> citizenship increasingly comes to be understood as shared rights, both human and civil, that all individuals should enjoy; as shared responsibilities of all human beings for survival of the planet, a clean environment, and a sustainable future; and as a collection of ethical principles and values that all humans embrace regardless of their cultural, ethnic, or religious backgrounds, the idea of citizenship that is shared and acknowledged by all humans is gaining strength.

The evolution of methods and programs used to train students for citizenship is a direct reflection of shifting societal perceptions of what role teachers and schools should play in the development of children's character. It is essential for educators to realize the fluid nature of educating students to become global citizens of character and begin exploring new methods to assist students in developing moral values and a sense of civic responsibility.

The use of film as an instructional tool continues to be a revolutionary resource because of its ability to provide students with visual representations of complex concepts in a global context. By empowering students to engage in critical examination, analysis, and discussion about issues of global character development depicted in films, teachers transform students from passive viewers into critical consumers of information. Students can begin looking at character lessons presented in popular films through their own cultural lens in order to gain a better understanding of their society, while also considering the global implications of these lessons. If a teacher shows a film addressing the character trait of patriotism, questions such as the following can help guide students to global character development.

- What does patriotism mean?
- How do I demonstrate my patriotism?
- How do people in other countries celebrate or demonstrate their patriotism?
- What are the similarities between demonstrating patriotism in my country compared to other countries?

Film is a unique instructional tool because it has the power to be a meaningful supplement to the classroom. Teaching students to critically examine and analyze films will enable them to transform from passive receivers of information into responsible consumers who are able to extract meaning from cinema. Incorporating film into classroom instruction to build citizens of character has two essential goals for students: helping them critically analyze and interpret films and engaging them in moral reflections to help them analyze their own values and beliefs in order to improve their decision making. Having an improved understanding of their own values and decision making will allow students to think deeply about what it means to be an effective citizen in such a diverse and ever-changing world. As many films will show, sometimes making the correct moral or civic decision is clouded by a variety of influences, situations, and consequences that individuals interpret differently. Allowing students to consider their own thoughts, feelings, attitudes, and beliefs through the engaging medium of film will empower them to rationalize and defend their own values, which is a necessary skill for all democratic citizens in the twenty-first century.

Note

For more information regarding this topic see Russell, W., & Waters, S. (2010). *Reel character education: A cinematic approach to character development.* Charlotte, NC: Information Age Publishing.

References

Cassata, D. (2012). US kids clamor for Congress to act against Kony. *Yahoo! News*. Retrieved from http://news.yahoo.com/us-kids-clamor-congress-act-against-kony-204739528.html

Driscoll, M. P. (2005). *Psychology of learning for instruction*. Needham Heights, MA: Allyn & Bacon.

Education Commission of the States. (2004). *Experts offer recommendations for improving citizenship education*. Retrieved from http://www.ecs.org/clearinghouse/54/18/5418.pdf

Lockwood, A. (2009). *The case for character education: A developmental approach*. New York: Teachers College Press.

Merryfield, M. (2011). Global education: Responding to a changing world. In W. Russell (Ed.), *Contemporary social studies: An essential reader* (pp. 57–77). Charlotte, NC: Information Age.

National Council for the Social Studies. (2005). *Position statement on global education*. Silver Spring, MD: Author.

Rapoport, A. (2012). The place of global citizenship in social studies curriculum. In W. Russell (Ed.), *Contemporary social studies: An essential reader* (pp. 77–97). Charlotte, NC: Information Age.

Russell, W. (2004). Teaching with film: A guide for social studies teachers. (ERIC Document Reproduction Service No. ED 530820)

Russell, W. (2007). *Using film in the social studies*. Lanham, MD: University Press of America.

Russell, W. (2012a). The art of teaching social studies with film. *The Clearing House, 85*(4), 157–164.

Russell, W. (2012b). The reel history of the world: Teaching world history with Hollywood films. *Social Education, 76*(1) 22–28.

Russell, W. (2012c). Teaching with film: A research study of secondary social studies teachers' use of film. *The Journal of Social Studies Education Research, 3*(1), 1–14.

Russell, W., & Waters, S. (2010). *Reel character education: A cinematic approach to character development*. Charlotte, NC: Information Age.

Salomon, G. (1994). *The interaction of media cognition and learning*. Hillsdale, NJ: Erlbaum.

United States Copyright Office. (1976). Title 17 (1) § 110 of the Copyright Law of the United States. Limitations on exclusive rights: Exemption of certain performances and displays. Retrieved from http://www.copyright.gov/title17/92chap1.html#110

CHAPTER ELEVEN

Meeting the Challenges of Implementing Global Education in a Time of Standardization

Mirynne Igualada
Dilys Schoorman

> This is great, but we will never be able to do global education in our classes. Not until we get rid of these tests.

This reaction from one of Schoorman's students, an hour into her first-ever graduate class in global education, has been seared into her consciousness as a teacher educator in Florida. It epitomizes the reality of teachers and students facing the accountability pressures unleashed by No Child Left Behind. It is particularly ironic for two reasons. Although the standardization movement and accountability pressures in education have been premised, ostensibly, on the need for the United States to be globally competitive, as teacher educators in different capacities, we have seen how these pressures are the catalyst for the opposite: an abandonment of global education. This reaction came in a graduate teacher education class in south Florida, home to the third largest population of immigrants and among the most diverse school districts in the nation. With the world in our classrooms, teachers felt threatened about engaging in global education.

This chapter emerged from the authors' efforts to step into this breach. Both authors are teacher educators in different contexts. Schoorman is a professor in multicultural and global education. Igualada—herself a former student in the aforementioned global education class (in a different year)—was a high school social studies teacher and currently works as an advanced studies curriculum coordinator. In her current role, she is able to share her work as a teacher with a wider group of teachers to demonstrate how teachers do not need to choose between test preparation or attention to standards and meaningful curriculum. This chapter draws on the insights of our experiences working with teachers in their struggles to engage in global education.

Rethinking Globalization

Even today, at least a decade into the ubiquitous use of the term "globalization," too many well-intentioned teachers are unaware that it is not synonymous with global education or internationalization. Whereas global education is an explicit study of issues and concerns that focuses on our interdependency as a planet, as highlighted by Hanvey's (1976) dimensions of global education, internationalization refers to understanding phenomena in terms of world context, rather than being confined to a local, national, insular perspective (Schoorman, 1999), globalization—as applied to education—signifies standardization with limited value for divergent perspectives or multiculturalism.

Critical educators have underscored the neoliberal, economic dimension of globalization (Apple, 2000; Giroux, 2004). They have noted its impact in education, where an extension of the values and practices of the unregulated hyper-capitalistic marketplace, has resulted in the corporatization of schools (Stromquist, 2002, 2007). The focus on efficiency, expediency, standardization, and, above all, consumerism and profiteering has turned public schools into captive markets for test makers. High-stakes accountability schemes that pit schools in competition for the scarce but badly needed funding, even as the payout of public education funds to testing companies has gone uncontested, have turned teachers into harried technicians in the classroom. The reframing of teacher effectiveness and student learning as isolated points on a single high-stakes test has reduced curriculum to test preparation drills. With emphasis placed on language arts and math, and teachers unable to be decision makers on meaningful curriculum, social studies teachers have been effectively silenced in any quest for transformatory curriculum development.

Scholars have argued, however, that the standardization pressures should not have to preclude social justice pedagogy (Bender-Slack & Raupach, 2008; Sleeter, 2005). There is support for how this might be done with resources ranging from organizations committed to social justice pedagogy such as Rethinking Schools (see http://www.rethinkingschools.org; http:// www.zinnedproject.org), advocates of critical pedagogy (see http:// www. Freireproject.com), and/or multicultural education (e.g., http:// www. nameorg.org; www.tolerance.org) and global education (see teachglobaled.net; see http://www.theglobalpeaceproject.com; Scott & O'Sullivan, 2002). Collectively, these scholars advocate an emancipatory perspective of global education aimed at "interrupting globalization" (Au & Apple, 2004, p. 784) through pedagogy that exemplifies counter-hegemonic praxis (Schoorman, 2000). Adopting the principles of critical pedagogy that facilitate dialogue and the reading of the world and the word (Freire, 1970/2000; 1998), critical global educators call for the analysis of power that views globalization "from below" (Bigelow & Peterson, 2002). They recommend the creation of a standards-conscious rather than standards-driven curriculum (Sleeter, 2005) that aspires to high levels of excellence measured in terms beyond test scores but that includes the knowledge, skills, and dispositions central to the development of citizens committed to social justice.

Challenges to Implementing Global Education Curriculum

The ideas presented in this chapter revolve around six challenges typically identified by teachers that hinder their engagement in global education curriculum:

- We will not be allowed to do this in our class. The administration expects a daily focus on test-based reading and writing skills.
- We do not have the time to do this.
- We do not have the knowledge to do this.
- This is not linked with the state standards.
- The students do not have the background knowledge for this.
- The students are not interested in learning about global topics.

In response to these concerns, we offer three sample lessons that can be adapted for use in multiple grade levels and that address state standards while also facilitating broad-based global awareness, citizenship skills, and critical thinking. The lessons were developed and implemented by Igualada in her various secondary social studies courses while she was enrolled as a graduate student in Schoorman's global education class. All three lessons could be used interchangeably to meet the six challenges.

Description of the Lessons

These three lessons have been selected because they explicitly highlight the manner in which 1) professional development (in this case, a graduate course) directly led to curriculum development in a teacher's classroom; 2) examples can address many of the concerns of teachers about the implementation of global education; and 3) critical pedagogy can be implemented relatively easily and effectively. In each case, a lesson meets content area as well as Common Core State Standards. Lesson 1 spans three to five class periods, Lesson 2 spans two to three class periods, and Lesson 3 can be taught in a single class period. Each lesson's effectiveness is discussed based on teacher observation and student comments as observed and recorded by Igualada.

Lesson 1: The Development Project. This lesson emerged from Igualada's own limited background knowledge and drew on her research on economic development lesson plans. She found a Web site with globalization lesson plans (see http://www. globalization101.org/teacher/development) and adapted the lesson plan to suit the students and subject area, focusing on the fundamental concepts of economic development and how it relates to globalization. In this lesson, students work collaboratively in a group writing a grant on behalf of a specific country of their choice that addresses one of the four strategies of development: poverty reduction, trade-not-aid, good governance, and sustainable development. The grant must include the following: 1) overview of the country and the general problems facing that country (political, social, economic, etc.); 2) introduction to the problem that will be addressed by the project; 3) description of the project in detail and how it will use the specific development strategy; 4) description of positive and negative impacts; 5) how grant money will be spent; 6) how success/failure will be measured; and 7) a detailed explanation of how the project will develop the country. This lesson plan clearly delineates what students need to know and do at a much higher cognitive level compared to state assessments, thus meeting and incorporating the basic skills that are driving current classroom instruction. Furthermore, the teacher is not expected to be an a priori expert on every country and its development issues; instead, expertise emerges as a consequence of the collective inquiry. This lesson combats the focus on standardized tests in the average classroom and the concern about the lack of background knowledge of teachers regarding global education.

Lesson 2: Ancient Ghana. Another frequent complaint regarding teaching global education is that students lack the background knowledge to make meaningful connections between the content area and their own frame of reference. In addition, teachers are under pressure to ensure that all lesson plans must be aligned to state or Common Core Standards. These concerns are addressed in a lesson on ancient Ghana, in which students work collaboratively through utilizing Project CRISS (Creating Independence through Student-owned Strategies). CRISS is a professional development tool that helps all content-area teachers increase students' literacy skills (see http://www.projectcriss.com). Each group is responsible for various sections from the chapter on the ancient kingdom of Ghana in the standard state-wide adopted textbook. These sections detail the rise and fall of Ghana, focusing on both economic and cultural achievements. Once this portion of the lesson is completed, students view the PBS *Frontline* special on digital dumping in Ghana (see http://www.pbs.org/frontlineworld/stories/ghana804/). Students watch the video; they are not given questions, nor are they required to take notes. Following the video students write an essay in which they reflect on what they learned from the video and compare/contrast ancient Ghana and Ghana today. They are prompted to reflect on what they think happened to Ghana between then and now and why Ghana is no longer a major power. In this lesson plan, the instructor is scaffolding the assignment with the ultimate goal of a writing assessment. The students are prepared to write at a high level through creating a well-researched and logical argument and are actively engaged through the process of creating and applying newly constructed knowledge to a current situation.

Lesson 3: Wealth Inequality. Many teachers also feel that their students are not interested in current global issues, and as result precious classroom time should not be spent on something that will not engage their students. An activity that contradicts this belief is from Bigelow and Peterson's (2002) *Rethinking Globalization*, "The Ten Chairs of Inequality" (pp. 115-117). Ten students are chosen to go sit in ten equally spaced chairs at the front of the room. The chairs represent the United States' wealth if it were divided equally. Students estimate how much wealth, in dollar amount, each chair represents. All of the students in the ten chairs are then asked to stand up while one student, selected to represent the richest 10 percent in the United States, occupies seven chairs; the remaining nine students then try to sit on the three remaining chairs. Through this activity students are provided a visual of how wealth is distributed. Once they understand—visually—how wealth is distributed, teachers can begin a conversation in which students are asked the following questions:

- What is it like, in real life, being a part of these three chairs? You can share personal experiences or those of others (without using names).
- What do those in power say to justify this distribution of wealth? What types of things have you heard or learned as you have gone through school?

These questions serve as a catalyst for engagement, as every student in the classroom has an opinion and experience regarding wealth.

Discussion of the Lessons

Each of these lessons is an illustration of how the six challenges typically encountered by teachers can be addressed with relative ease through the use of existing instructional resources available on the Internet. Teachers' concerns about the lack of time and the perception that global education is added work in an already packed curriculum are addressed through the demonstration of how such discussions can be integrated into existing or mandated curriculum (e.g., the lesson on Ghana) or with relatively short but powerful lessons (e.g., the lesson on inequality). We make the case for the development of similar global education lessons by highlighting four observations related to their positive impact on student engagement, teacher empowerment, academic skills, and the transformatory potential of social studies itself.

Global Education and Enhanced Student Engagement

What has been clear to Igualada, as a classroom teacher, and several of her colleagues in a graduate class on global education who were in-service teachers, was that lessons in global education have the capacity to facilitate high levels of student engagement. Student engagement was demonstrated through activities that required students to gather their own data and make a case for a nation of their choice (Lesson 1), to engage in reflective thinking about the interconnections between past and present, the impact of macro concerns (such as globalization and colonization) on micro-level community experiences (Lessons 1 and 2) to raise critical questions grounded in concerns for social justice (Lesson 3), and to begin to engage their skills as advocates for nations around the world (Lessons 1 and 3). Not only do these lessons combat the complaint that students do not care about global issues, they also facilitate a student-centered classroom environment.

There were several indicators of student engagement in just these three lessons. Foremost were the student comments and questions raised during the lessons. Among the clearest indications of student engagement were the following comments that were made during small-group or whole-class discussion. These comments are representative of the critical insightfulness that the students express while working in peer groups:

- I think good governance might have something to do with law and maybe there needs to be better law schools and teachers.
- Why can't girls go to school?

- My country only grows one thing to export; I think that they need to grow more to have food for their people.
- Do you think our computers went there [landfills in Ghana] when we got new ones in the media center?
- What happens to our stuff when we put it out?
- I don't even know where batteries go.

Furthermore, the activities of the lessons required that students engage in problem solving, division of tasks, and discussion of their ideas. Student engagement was also cross-curricular in nature. The majority of the students were also enrolled in environmental science, and the lessons resulted in discussions on pollution as students began to make connections regarding materialism and consumption as they relate to unequal power structures in the past and present, wealth inequality, and economic development. Allowing students to generate and drive discourse demonstrated that they would continue to think and discuss after they left the class.

Global Education: An Opportunity for Teacher Empowerment

In a context in which most teachers follow a standardized curriculum that fosters a "drill-and-kill" approach to teaching and learning, and in which social studies has been relegated to topics inserted into reading, writing, or mathematics exercises, the development of a global curriculum offered Igualada the opportunity to reclaim her role as a curriculum decision maker. Perhaps the most salient fact about the lessons presented here is that the teacher could, with relative ease, draw on existing online resources to develop lessons of global significance. The lessons could be integrated into an already existing curriculum (e.g., Lesson 2) or could be presented in very short episodes (Lesson 3) during which minimal class time was taken up for the planting of conceptual seeds that will have long-term relevance. These lessons also facilitated a change in the traditional student–teacher dynamic, energizing the instructional process. Teaching was no longer defined as the instructor delivering content; instead, the process of tackling a real-world problem allowed the teacher to simultaneously become the student. Significant in this shift in the teacher's role was Igualada's continued ability to integrate the prescribed standards with her choice of subject matter, demonstrating that there was space for teachers to engage in innovative curriculum development despite what is perceived as an oppressive regime of accountability and standardization.

Global Education: High Standards for Academic Skills

The lessons presented here addressed a wide variety of standards, demonstrating that a standards-conscious curriculum is not antithetical to high-quality learning. In these lessons students must read a variety of sources, analyze charts, understand advanced vocabulary, compare and contrast strategies, and recognize bias, all of which meet standards and skills required for state assessments. They require students to research background knowledge of historical events and current political relationships. Through the classroom discussion that these lessons promote, students are able to think critically about the issue through listening to others' viewpoints and questioning their rationale and/or bias. Students practice their oral communication skills by relaying personal experiences and engaging in thoughtful discourse. The lessons are concrete examples of district and state requirements of complex tests, essay writing, and critical thinking, underscoring the fact that these skills need not come from a worksheet. Rather, these skills can be fostered through solving real-world problems in a group setting where students collaborate, communicate, and create a finished product.

These activities also promote social and intellectual skills that are important for college and workforce readiness and are not included in the standardization movement. For example, students must figure out how to divide the workload, which requires collaboration. Students must also write and present, promoting oral and written fluency. Furthermore, students are required to access and analyze information from a variety of sources and come up with creative and innovative solutions to problems. All of these skills are at the center of district-level discourse.

Global Education: An Opportunity for Counter-Hegemonic Praxis

These lessons exemplify efforts that counter the "top-down" neoliberal tendencies of globalization and demonstrate how global education, grounded in critical pedagogy and social justice, might serve as a foundation for the development of global citizenship among students. A first step in this quest is making students aware of the inequalities that exist within the world and their own (privileged) positions in these economic, political, and social relationships. These lessons also demonstrate efforts to engage youth in critical consciousness-raising, often deemed too difficult, advanced, or controversial for a public school classroom.

As students are required to understand complex political, social, and economic components that make up these lessons, they begin to question the existing power structures that exist both at home and abroad. For example, in Lesson 1, a student will notice that the literacy rate for females is much lower and begin to research why this has occurred. This thought process usually leads to an understanding that poverty is linked to the status of females in a society. Students revealed their propensity to examine their own actions and blind spots (e.g., acknowledging not knowing where their batteries or electronic trash ended up), and the transcontinental interconnectedness of those actions (e.g., wondering whether their old school computers ended up in Ghana's landfills), and to advocate on behalf of people about whom they had just learned. Students' questions revealed a concern for humanity at a global level that could potentially have a profound effect even on the adults around them.

Implications for the Classroom

If teachers are to successfully engage in counter-globalization efforts, they must be supported in these efforts. This support needs to come from 1) teacher education programs, where they actually experience emancipatory pedagogy, and 2) readily available resources for curriculum development. This does not mean that teachers should be provided "recipes" for blind implementation in the classroom; on the contrary, as demonstrated in these examples, existing reliable resources served as a starting point for adaptation. Curriculum for critical consciousness-raising cannot be scripted; what teachers need are materials and resources that serve as catalysts for student engagement and social justice praxis.

The more challenging implication is the reframing of the role of teachers as critically engaged practitioners in counter-hegemonic praxis. In contexts (including teacher education programs) where teaching has succumbed to being framed in terms of compliance with external accountability regimes (even when such compliance results in increasing injustice), this reframing is more difficult, but also more crucial. We can ill afford to allow teachers and teacher educators to participate in their own silencing and de-professionalization. These lessons offer a glimpse into how this trend can be countered, even if in modest ways, one teacher, one classroom, one program at a time.

We also need to recognize the tremendously important role that social studies plays in the individual and social empowerment of students and their communities, an approach to education that is largely absent in current educational practices, especially those subject to compliance with standardization and accountability mandates. As educators such as Howard Zinn (2007) and

James Banks (2001) have noted, social studies, as a field, has a tremendous potential for facilitating civic and social activism against social injustice; however, social studies teachers must be able to seize the limited opportunities within the current curricular structures, in which they can engage in counter-hegemonic pedagogy. The lessons presented here alert teachers to these possibilities.

Finally, we live in a time when global education should no longer be a curricular choice just for privileged students. All students need to be educated about global issues and now more explicitly about the impact of globalization. The failure to examine globalization from a critical perspective will only do further damage to students already marginalized by current educational policies and practices. The lessons presented here offer teachers examples that are both pragmatic and conceptually rigorous in alerting students to issues of social injustice on a global scale, and the manner in which they are all intertwined in these systemic processes.

References

Apple, M. (2000). Between neoliberalism and neoconservatism: Education and conservatism in a global age. In N. Burbules & C. Torres (Eds.), *Globalization and education: Critical perspectives* (pp. 57–78). New York: Routledge.

Apple, M., Au, W., & Gandon, L. A. (Eds.). (2009). Mapping critical education. In *The Routledge international handbook of critical education* (pp. 3–19). New York: Routledge.

Au, W. (2007). High-stakes testing and curricular control: A qualitative metasynthesis. *Educational Researcher*, 36(5), 258–267.

Au, W., & Apple, M. (2004). Interrupting globalization as an educational practice. *Educational Policy*, 18(5), 784–793.

Banks, J. (2001). Teaching social studies for decision making and citizen action. In C. Grant & M. L. Gomez (Eds.), *Campus and classroom: Making schooling multicultural* (2nd ed., pp. 109–134). Upper Saddle River, NJ: Merrill/Prentice Hall.

Bender-Slack, D., & Raupach, M. P. (2008). Negotiating standards and social justice in the social studies: Educators' perspectives. *The Social Studies*, 99(6), 255–259.

Bigelow, B., & Peterson, B. (Eds.). (2002). *Rethinking globalization: Teaching for justice in an unjust world*. Milwaukee, WI: Rethinking Schools.

Freire, P. (1970/2000). *Pedagogy of the oppressed*. 30th anniversary edition. New York: Continuum.

Freire, P. (1998). *Teachers as cultural workers: Letters to those who dare to teach*. Boulder, CO: Westview Press.

Giroux, H. A. (2004). *The terror of neoliberalism: Authoritarianism and the eclipse of democracy*. St. Paul, MN: Paradigm.

Hanvey, R. (1976). *An attainable global perspective*. New York: Center for Global Perspectives in Education.

Jones, B. D., & Egley, R. J. (2007). Learning to take tests or learning for understanding? Teachers' beliefs about test-based accountability. *The Education Forum*, 71(3), 232–248.

Schoorman, D. (1999). The pedagogical implications of diverse conceptualizations of internationalization: A US-based case study. *Journal of Studies in International Education*, 3(2), 19-46.

Schoorman, D. (2000). What really do we mean by internationalization? *Contemporary Education*, 71(4), 5-11.

Scott, T. J., & O'Sullivan, M. (2002). Essential websites to research the globalization process. *The Social Studies*, 93(5), 232-236.

Sleeter, C. (2005). *Un-standardizing curriculum: Multicultural teaching in standards-based classrooms*. New York: Teachers College Press.

Sleeter, C. (2008). Equity, democracy, and neoliberal assaults on teacher education. *Teaching and Teacher Education*, 24(8), 1947-1957.

Stromquist, N. P. (2002). *Education in a globalized world: The connectivity of economic power, technology and knowledge*. Mahwah, NJ: Rowman & Littlefield.

Stromquist, N. (2007). Internationalization as a response to globalization: Radical shifts in university environments. *Higher Education*, 53(1), 81-105.

Zinn, H. (2007). Why students should study history. In W. Au, B. Bigelow, & S. Karp (Eds.), *Rethinking our classrooms* (Vol. 1, pp. 179-185). Milwaukee, WI: Rethinking Schools.

CHAPTER TWELVE

Definition Devolution: Allowing Students to Redefine and Rename Citizenship and Civic Engagement

Emma K. Humphries
Elizabeth Yeager Washington

It should come as no surprise that the diverse meanings and understandings of the terms "citizenship" and "civic engagement" make it difficult to prepare today's students to assume their roles as citizens in a politically and technologically interdependent world. This chapter examines a conceptual framework through which educators can help students to understand these terms from a twenty-first-century perspective that takes globalization and technological innovation into account. We will review some of the existing research related to our topic and summarize a variety of views and perspectives that have clear implications for citizenship/civic engagement. Finally, we will integrate these different areas of research in order to construct our own instructional strategy that allows students to create, redefine, and rename the terms "citizenship" and "civic engagement" in a way that has personal meaning for them.

Diverse Perspectives on Twenty-First-Century Citizenship and Civic Engagement

As Westheimer and Kahne (2004) note, definitions of citizenship have been and will likely continue to be debated. Unfortunately, they explain, the most narrow and traditional definitions are the ones typically found in dictionaries and social studies texts. These definitions tend to be limited to what Russell Dalton (2009) calls "duty-based citizenship," which encompasses "the formal obligations, responsibilities, and rights of citizenship" (p. 5). In addition to being particularizing and exclusionary, these views of citizenship tend to neglect twenty-first-century trends and developments (Cohen, 1999). This is problematic, as traditional definitions for citizenship (those regarding the status of a citizen with rights and duties) and even for civic engagement (individual and collective actions designed to identify and address issues of public concern) tend to be "old-fashioned, primly moralistic, and limiting" (Levine,

2007, p. 1) and may no longer be viable or meaningful to our students in an increasingly global and social media-saturated socio-political landscape (Cohen, 1999; Jenkins, 2006; Steward, 1991; Sunal, 2008). Worse still, these two terms may come across as abstract or irrelevant. As teacher educators we may find that we provide students with definitions for these concepts, but in the process, we effectively exclude those who do not fit the traditional notions.

The literature on democratic citizenship addresses the complexity of defining the concept of democracy and active citizenship, both in theory and in practice. Scholarship in democratic citizenship education has long emphasized teaching students about living in and contributing to a democratic society. Dewey (1916, 1927) argued that the school should be a democracy in microcosm, where pupils learn particular processes, values, and attitudes to live effectively as citizens. Democracy, to Dewey, largely meant a form of active community life—a way of being and living with others. Moreover, Dewey emphasized that democracy entails certain habits of mind that must be cultivated throughout citizens' lives as they participate in various institutions and groups through which they have a voice in setting goals, sharing knowledge, communicating, and taking direct action. Most important, Dewey envisioned democracy as a creative and constructive process for which citizens needed practical judgment, a shared fund of civic knowledge, and deliberative skills and dispositions—much of which must be learned in schools.

Nearly eighty years later, Parker (1996a, 1996b) continued to raise important questions about what it means to educate children "for the demands of an increasingly diverse society that is struggling to realize the democratic ideal" (1996b, p. 2). Instead of citizenship education forging a homogenized national identity, he argues, it should prepare students to live in a pluralistic society that allows for a wide range of cultural and ethnic identities. In other words, we must strive for "democratic political community within cultural pluralism" (1996b, p. 20). For Parker, the central citizenship question is, "How can we live together justly, in ways that are mutually satisfying, and that leave our differences, both individual and group, intact and our multiple identities recognized?" (1996a, p. 113).

Parker asserts that the school's "first moral obligation" is to give children an education that will equip them to take advantage of their citizenship (1996b, p. 2). Indeed, in Parker's view, schools already possess the "bedrocks of democratic living—diversity and mutuality" (1996b, pp. 2, 10). His conception of democratic citizenship education values direct involvement in public life, pluralism, and democracy as a way of life involving "deliberation, action, and reflection" (p. 121). He argues for a discourse of "responsibility, negotia-

Definition Devolution

tion, and obligation" aimed at creating a "broad political comradeship" (creating the political "one" out of the cultural "many") (p. 117).

Another conception of democratic citizenship education that informs our framework is that of "critical democracy." Critical democracy implies a moral commitment to place the public good over individual power and privilege (Barber, 1984; Dahl, 1982; Gran, 1983). It also implies an effort by citizens to address meanings of deliberation, civic responsibility, social equity, group conflict and cooperation, community, individual rights, institutional organization, public interest, and the distribution of power (Barber, 1984; Ventriss, 1985). "Critical democratic citizenship" comprises the exercising of skills of critical inquiry and analysis to help make meaning of what is happening in civic relationships and institutions in the world around us. As we ask questions and seek knowledge, we can use our understanding to challenge existing power structures.

Dewey (1916, 1927) and Parker (1996a, 1996b, 2008), as well as C. Wright Mills (1956), conceptualized civic engagement as active participation in civic institutions in order to influence governance. Parker (2008) describes this behavior as "enlightened political engagement" or "wise political action" (p. 68), explaining that

> political engagement refers to the action or participation dimension of democratic citizenship, from voting to campaigning, boycotting, and protesting. Democratic enlightenment refers to the knowledge and commitments that inform this engagement: for example, knowledge of the ideals of democratic living, the ability to discern just from unjust laws and action . . . and the ability and commitment to deliberate public policy in cooperation with disagreeable others. Without democratic enlightenment, participation cannot be trusted . . . [and] can be worse than apathy. (p. 68)

Parker reminds us that enlightened political engagement is not easy to accomplish; in fact, it is a continuous goal toward which we work with others who hold ideas and perspectives different from those that we hold.

> citizenship is "not just an attitude of mind or even a subject of political education," it is rather a set of entitlements common to all members of society. . . . I like to think of citizenship as a set of chances—life chances—that define a free society. [This] involves basic rights, equality before the law, due process, the integrity of the person, freedom of expression and association. It also involves chances of participation, universal suffrage, of course, but equally importantly market access including labour market access, and social movement in the numerous opportunities of civil society. This is what citizenship means in the full sense of the word. . . . [Citizenship] provides an instrument for living with difference with regard to how people act with and toward other citizens, societies and cultures within a global community. (pp. 62-63)

Jones and Gaventa (2002) specifically conceptualize active citizenship as "the direct ways in which citizens influence and exercise control in governance" and "the direct intervention of citizens in public activities," as well as the "accountability of the state and other responsible institutions to citizens" (p. 7). This "relational dynamic" of citizenship places "obligations on both citizens and the state through participatory democratic systems . . . [that] require direct connection between citizens and the state" (p. 7). This, in turn, "entails institutional reforms that enable democratic participation through the production of new forms of relationship between civil society and the state" (p. 7). They conclude, "When citizens perceive themselves as actors in governance, rather than passive beneficiaries of services and policy, they may be more able to assert their citizenship through actively seeking greater accountability . . . and shaping policies that affect their lives" (p. 7).

By contrast, political scientists and sociologists tend to differentiate between political and civic engagement. In *A New Engagement?*, Zukin, Keeter, Andolina, Jenkins, and Delli Carpini (2006) describe political engagement as "activity aimed at influencing government policy or affecting the selection of public officials," and civic engagement as "participation aimed at achieving a public good, but usually through direct hands-on work in cooperation with others" (p. 51). They explain: "Civic engagement normally occurs within nongovernmental organizations and rarely touches upon electoral politics" (p. 51). They refer to volunteering in one's community as the most obvious example. Their view is similar to Henry Brady's (1999) distinction between electoral activities (voting and campaign activity) and nonelectoral activities, which he further delineates as either "conventional" (participating in informal community work, attending meetings, being an organizational member) or "unconventional" (signing petitions or participating in demonstrations or boycotts). In *Bowling Alone*, Robert Putnam (2000) distinguishes between what he calls "expressive" forms of behavior (voting, writing letters, or discussing political affairs), and "cooperative" activities (such as working to improve a community problem). Peter Levine (2007) operationalizes civic engagement as a list of variables across three categories: community participation, political engagement, and political voice.

The literature in this area is replete with useful national and cross-national studies that demonstrate the variety of ways in which active citizenship can be conceptualized (e.g., Davies, 2006; Davies & Issitt, 2005; Ibrahim, 2005; Jones & Gaventa, 2002; Torney-Purta, Lehmann, Oswald, & Schulz, 2001; Torney-Purta & Richardson, 2004; Watts, 2006). While helpful, these diverse meanings and understandings of citizenship and civic engagement still tend to neglect globalization and technological innovation. Before we challenge our students to redefine and rename these terms, it is instructive for us to consider alternative notions ourselves.

For its recognition of globalization, Cogan and Derricott's (1998) idea of multidimensional citizenship is informative. It includes eight key characteristics with implications for being active as a global citizen: 1) the ability to look at and approach problems as a member of a global community; 2) the ability to work with others in a cooperative way and to take responsibility for one's role/duties in society; 3) the ability to understand, accept, appreciate, and tolerate cultural differences; 4) the capacity to think in a critical and systematic way; 5) the willingness to resolve conflict in a nonviolent manner; 6) the willingness to change one's lifestyle and consumption habits to protect the environment; 7) the ability to be sensitive toward and to defend human rights; and 8) the willingness and ability to participate in politics at local, national, and international levels. They conclude that

> [the task is] to help [future] citizens recognize the global challenges which affect each of us personally and are part of our individual and social responsibility to address. Put simply, global challenges cannot be left for someone else to deal with: rather, the responsibility lies with each of us to safeguard global well-being. . . . Twenty-first century citizenship will require active citizenship participation—citizens who view themselves as actors in the world. (p. 133)

Likewise, Oxfam International conceptualizes a Global Citizen as someone who is aware of the wider world and has a sense of one's own role as a world citizen; respects and values diversity; has an understanding of how the world works economically, politically, socially, culturally, technologically, and environmentally; is outraged by social injustice; participates in and contributes to the community at a range of levels from local to global; is willing to act to make the world a more sustainable place; and takes responsibility for his or her actions (http://www.oxfam.org/en). In this way, global citizenship transcends legal conceptions and the knowledge that we are citizens of the globe and moves us toward an acknowledgment of our responsibilities both to each other and to the Earth itself.

Bennett (2008), in offering an alternative to the traditional civic education ideal of the "dutiful citizen," outlines the "actualizing citizen" model, which includes global awareness and social-networking technologies. While the dutiful citizen votes, participates in government-centered activities, follows the mass media, and joins civic organizations, the actualizing citizen has a higher sense of individual purpose, engages in personally defined acts (e.g., consumerism, volunteering, transnational activism), mistrusts mass media, and favors loose networks of community action, which are "often established and sustained through friendships and peer relations and thin social ties maintained by interactive information technologies" (p. 14).

Zukin and colleagues (2006) focus on the concept of "lived citizenship," which offers an alternative way of thinking about youth citizenship and civic engagement as a young person's embodied perspective and lived experience. As they explain, lived citizenship from the perspective of the young person is

> experiencing oneself as having to do something on a specific issue or condition because it is the right thing to do and not to do so would go against who I am, and when I work with others on this issue, I experience myself as able to make a difference, and I feel good about this and about myself. (p. 12)

While this conception mentions neither technology nor globalization, it leaves room for both by avoiding the narrow list of behaviors we find in traditional definitions.

A Five-Step Lesson Plan for Redefining and Renaming Citizenship and Civic Engagement

Given these diverse meanings and understandings of citizenship and civic engagement, we suggest that young people be afforded opportunities to think about and perhaps even reclaim them for their generation. Accordingly, the five-step instructional process outlined below has three key objectives: to challenge students to question traditional notions and definitions of citizenship and civic engagement; to allow students to redefine and rename these terms in a way that acknowledges globalization, embraces twenty-first-century technologies, and holds meaning for them; and to encourage students to construct personal civic identities that consider the new definitions and names. We suggest that this plan can be modified for either middle or high school simply by adjusting the reading level of the sources and the scope of the research students do.

Step One: Questioning Citizenship and Civic Engagement. As a first step, students could examine traditional notions and definitions of citizenship and civic engagement, and consider ways in which these terms might be insufficient today, in light of globalization and the ever-changing technologies that tremendously affect both our daily lives and the larger functioning of our society. Specifically, students could critically evaluate the ways in which traditional notions and definitions might be exclusionary and outdated.

Depending on classroom resources and student skill levels, teachers could approach this process in a variety of ways. We suggest that the teacher display two questions—What is citizenship? and What is civic engagement?—on the front board and provide print and/or Web sources for students to work in pairs to answer the questions. Including the class textbook as a source would also be useful. At this point it is essential that these resources point students to more traditional notions of citizenship and engagement.

In order to answer the two questions, students should be instructed to look for definitions for the terms, provide a citation for each definition, and identify five words that recur and/or seem important. When they have finished this task, the teacher can use Wordle.com to create and display a word cloud using the words that students have identified. At this point, the teacher can facilitate a discussion around questions such as the following:

- What seems right about these definitions?
- What seems wrong about them?
- Are they lacking in any way? How?
- Do they make sense to you? Why or why not?

Students should be encouraged to consider the increasingly global society in which we live and the impact of technology on democratic processes.

Step Two: Redefining and Renaming Citizenship. As a second step, students should research current notions and definitions for citizenship, write their own one- to three-sentence definition for the term, and give it a new name. As with the previous step, there are various ways to approach this. We suggest that students continue to work in pairs, with a computer for each pair if possible. Before students begin their research, the teacher can lead a whole-class brainstorming session on potentially effective search terms (e.g., Modern Citizenship, Citizenship + Technology, Globalization + Citizenship, etc.). The teacher can keep a running list on the front board and instruct students to begin working through the list in their search.

At this point, students should again identify what they consider to be the top three notions or definitions, provide a citation for each, and identify words that recur and/or seem important. The teacher should emphasize that students get to decide which notions, definitions, and words are best. After recording their top three choices and compiling their list of important words, the teacher can ask students to again write their own one- to three-sentence definition for citizenship and give it a new name that is one to three words long. When students have completed this task, the teacher can allow time for pairs to share their new definitions and names.

Step Three: Assigning Action. For step three, students should make a list of ten activities that they would characterize as important for good citizens. However, before they begin working on their lists, the entire class should discuss traditional activities that are often mentioned in textbooks or other common or mainstream explanations of participatory citizenship. Examples might include voting, writing a letter to one's member of Congress, volunteering, and working to solve a community problem. The class should discuss the following:

- Which of these activities still hold value today? Why?
- Which seem outdated or useless? Why?
- Which may still be useful, but are perhaps done differently because of the impact of globalization and technology?

This naturally leads students to begin working on their list of ten activities in which citizens engage today. They should be encouraged to be as specific as possible in order to account for and to include the factors of globalization and technology in their lists. For example, does a motivated citizen today tweet on Twitter instead of sending a snail-mail letter to her member of Congress to voice her outrage about proposed legislation (e.g., Stop Online Privacy Act)? Does he write a blog post instead of sending a letter to the editor of a newspaper when he is frustrated by the actions of a nonprofit (e.g., Susan G. Komen pulling funding from Planned Parenthood)? Does he change his profile picture on Facebook instead of attending a public protest to express his anger over the actions of law enforcement (e.g., the Trayvon Martin case)? Or does she use various social media sites instead of traditional advertising media to broadcast information about an antigovernment demonstration (e.g., Arab Spring)? Do these new types of engagement make a difference or have value? Why or why not? As with the previous step, once students have completed this task, the teacher should allow time for pairs to share some activities from their lists.

Step Four: Redefining and Renaming Civic Engagement. As a fourth step, students should repeat step two, only this time for civic engagement. That is, they should research more current notions and definitions for civic engagement, write a one- to three-sentence definition for the term, and give it a new name. Again, students should work in pairs, with one computer for each pair, and then share their definitions and new names with the class.

Step Five: Building a Civic Identity. During the fifth and final step, students can work either in pairs or individually to create a Citizen Identity Profile (CAP) that incorporates their definitions for citizenship and civic engagement. Each CAP should include an avatar, a list of Civic Superpowers (skills and talents that make up effective citizenship), a list of causes that are important to the students, and a list of civic activities in which the students would like to engage. If technology is available, students can use a digital avatar creator, but it would be equally meaningful for students to draw their avatars and write out their lists of superpowers, causes, and activities.

Conclusions

Of course, the lessons of citizenship and civic engagement are best learned through actual civic engagement itself. Nonetheless, we believe that affording instructional time for the thoughtful and deliberate reconceptualization and redefinition of these terms holds tremendous import for student civic identity. To that effect, in helping students to find meaning for citizenship and civic engagement, it is useful to take the broadest view possible. To do otherwise would be to potentially exclude the students themselves. VeLure Roholt, Hildreth, and Baizerman (2008) argue that society falsely accuses young people of being unengaged, because 1) the notion of "citizen" is often reserved for certain types of engagement and not others, 2) some types of civic engagement are age graded and therefore not open to young people, and 3) adults do not perceive your involvement in certain non-age-graded activities as what it means to do and be a citizen.

Indeed, Bennett (2008) states that young people have been falsely dichotomized as either reasonably active and engaged or relatively passive and disengaged. The first label "emphasizes generational changes in social identity that have resulted in the growing importance of peer networks and online communities" (p. 2). The second, while it includes the rise in alternative forms of expression (e.g., consumer politics, protests on social network sites), assumes that the decline in traditional forms of engagement (e.g., voting, following public affairs in the news) should be lamented. As teacher educators, we would serve our students well to embrace the engaged youth paradigm while also creating opportunities to think about it in a completely different way that takes globalization and technology into account.

References

Barber, B. (1984). *Strong democracy: Participatory politics for a new age.* Berkeley: University of California Press.

Bennett, W. L. (2008). Introduction. *Civic life online: Learning how digital media can engage youth* (pp. 1-24). Cambridge, MA: MIT Press.

Brady, H. (1999). Political participation. In J. P. Robinson, P. R. Shaver, & L. S. Wrightsman (Eds.), *Measures of political attitudes* (pp. 65-80). New York: Academic Press.

Cogan, J. J., & Derricott, R. (1998). *Citizenship for the 21st century: An international perspective on education.* London: Kogan Page.

Cohen, J. L. (1999). Changing paradigms of citizenship and the exclusiveness of the demos. *International Sociology,* 14(3), 245-278.

Dahl, R. (1982). *Dilemmas of pluralist democracy.* New Haven, CT: Yale University Press.

Dahrendorf, R. (1997). *After 1989: Morals, revolution and civil society.* New York: Palgrave Macmillan.

Dalton, R. L. (2009). *The good citizen: How a younger generation is reshaping American politics.* Washington, DC: CQ Press.

Davies, L. (2006). Global citizenship: Abstraction or framework for action? *Educational Review,* 58, 5-25.

Davies, I., & Issitt, J. (2005). Reflections on citizenship education in Australia, England, and Canada. *Comparative Education,* 41, 389-410.

Dewey, J. (1916). *Democracy and education.* New York: Macmillan.

Dewey, J. (1927). *The public and its problems.* Chicago: Swallow.

Gran, G. (1983). *Development by people: Citizen construction of a just world.* New York: Praeger.

Ibrahim, T. (2005). Global citizenship education: Mainstreaming the curriculum. *Cambridge Journal of Education,* 35, 177-194.

Jenkins, H. (2006). *Convergence culture: Where old media and new media collide.* New York: New York University Press.

Jones, E., & Gaventa, J. (2002). *IDS development bibliography 19: Concepts of citizenship.* Brighton, England: Institute of Development Studies.

Levine, P. (2007). *The future of democracy: Developing the next generation of American citizens.* Medford, MA: Tufts University Press.

Mills, C. W. (1956). *The power elite.* New York: Oxford University Press.

Parker, W. C. (1996a). "Advanced" ideas about democracy: Toward a pluralist conception of citizen education. *Teachers College Record,* 98, 104-125.

Parker, W. C. (1996b). Introduction. *Educating the democratic mind* (pp. 1-22). Albany, NY: SUNY Press.

Parker, W. C. (2008). Knowing and doing in democratic citizenship education. In L. S. Levstik & C. A. Tyson (Eds.), *Handbook of research in social studies education* (pp. 65-80). New York: Routledge.

Putnam, R. D. (2000). *Bowling alone: The collapse and revival of American community.* New York: Simon & Schuster.

Steward, F. (1991). Citizens of planet Earth. In G. Andrews (Ed.), *Citizenship,* (pp. 67-75). London: Lawrence & Wishart.

Sunal, C. S. (2008). What is a citizen? The impact of the Internet on definitions of citizenship. In P. J. Vanfossen & M. J. Berson (Eds.), *The electronic republic? The impact of technology on education for citizenship* (pp. 17-36). West Lafayette, IN: Purdue University Press.

Torney-Purta, J., Lehmann, R., Oswald, H., & Schulz, W. (2001). *Citizenship and education in twenty-eight countries: Civic knowledge and engagement at age fourteen.* Amsterdam: International Association for the Evaluation of Educational Achievement. Retrieved from http://www.wam.umd.edu/~iea

Torney-Purta, J., & Richardson, W. (2004). Anticipated political engagement among adolescents in Australia, England, Norway, and the United States. In J. Demaine (Ed.), *Citizenship and political education today* (pp. 41–58). Basingstoke, England: Palgrave Macmillan.

VeLure Roholt, R., Hildreth, R. W., & Baizerman, M. (2008). *Becoming citizens: Deepening the craft of youth civic engagement.* New York: Haworth.

Ventriss, C. (1985). Emerging perspective on citizen participation. *Public Administration Review, 47*, 433–440.

Watts, M. (2006). Citizenship education revisited: Policy, participation, and problems. *Pedagogy, Culture, & Society, 14*, 83–97.

Westheimer, J., & Kahne, J. (2004). What kind of citizen? The politics of educating for democracy. *American Educational Research Journal, 41*(2), 237–269.

Zukin, C., Keeter, S., Andolina, M., Jenkins, K., & Delli Carpini, M. X. (2006). *A new engagement?,* New York: Oxford University Press.

CHAPTER THIRTEEN

Hearing a Chorus of Voices: Globalizing the U.S. History Curriculum with Historical Empathy

Joseph O'Brien
Jason L. Endacott

> As citizens of the global community, students also must develop a deep understanding of the need to take action and make decisions to help solve the world's difficult problems. They need to participate in ways that will enhance democracy and promote equality and social justice in their cultural communities, nations, and regions, and in the world.
>
> J. A. Banks, 2008, pp. 134–135

The changing face of American society compels us to diversify and globalize the U.S. history curriculum beginning with the meaningful inclusion of a wider range of historical figures and the promotion of historical inquiry that incorporates historical empathy as a means for students to capture these figures' "voices." The typical K–12 history curriculum in the United States emphasizes a common national heritage over pluralistic history, which runs the risk of promoting a "discourse of invisibility . . . true of every non-European group of people who constitute our nation" and portraying history as "an incoherent, disjointed picture of those who are not White" (Ladson-Billings, 2003, p. 4). State history standards typically view history of certain peoples, such as African Americans or immigrant groups, through the lens of the U.S. government or through their interaction with the government. In turn, history textbooks over time have placed differing levels of importance on the historical experiences of these groups depending upon the significance of their interactions with the federal government. For example, after analyzing U.S history textbooks published in the mid-twentieth century, FitzGerald (1979) concluded that, "Blacks . . . were quite literally invisible" (p. 84), while later textbooks published after the many African American accomplishments of the civil rights movement found them "moved to the center stage of American history" (Lerner, Nagai, & Rothman, 1995, p. 71).

The resulting effect on minority and immigrant groups in the K–12 American history curriculum is one of shifting prominence, marginalization, and irrelevance while their experiences and achievements are often depicted only to the extent that they reinforce the notion of national achievement (Barton, 2009). Further compounding this problem is the manner in which non-white experiences are typically conveyed through government documents such as court cases and congressional acts. Reliance upon these documents has been demonstrated to be problematic because they lack voice (Paxton, 1999, 2002) and cast the impression that minority and immigrant peoples are simply acted upon by the government, thereby robbing them of their historical voice, wisdom, and uniqueness as a people. The resulting impression of U.S. history is one in which the U.S. government, rather than people, is the nexus of change.

Beyond the problems of exclusion and lack of agency, American students are also rarely exposed to historical viewpoints that originate from outside the United States (e.g., Lindaman & Ward, 2006). This narrow national-centric focus prevents students from realizing the global context in which U.S. history has occurred, not to mention benefiting from insight that the voices of those who lived beyond U.S. borders but were intimately affected by its actions can provide. Failing to globalize people's voices serves to reinforce an American exceptionalism approach to the U.S. history curriculum, one that portrays the American experience as more significant or of greater value than those from around the world.

It is therefore crucial that American students are exposed to the lived experiences and voices of a much broader range of historical figures both within the United States and abroad. However, simply adding a plurality of American and global voices is not enough to avoid these risks. If our students fail to contextualize and appreciate the lived experiences of the people who raise them, then we further risk marginalizing knowledge about such individuals (King, 1995) and reducing them to optional sidebars in the margins of textbooks that leave the "monocultural, exclusive narrative undisturbed" (Ladson-Billings, 2003, p. 9). In short, our students need not only access to, and emphasis on, the voices of all manner of global historical figures but also the means by which they can understand their lived experiences and contextualize them into a broader understanding of history. Historical empathy, the "complex balance between considering the perspectives of and connecting with people in the past" (Kohlmeier, 2006, p. 37), offers a means to give life to those voices because it allows students to build an affective bridge to the lived experiences of historical figures through similar shared experiences (Endacott, 2010).

Historical empathy is the process of students' holistic engagement with historical figures to better understand and contextualize their lived experiences, decisions, or actions. Historical empathy helps us understand how people from the past thought, felt, made decisions, acted, and faced consequences within a specific historical and social context. When students engage in historical empathy as a tool of historical inquiry they delve deeply into the context of the past through the use of primary and secondary source evidence to explore the values, beliefs, experiences, and decisions of historical figures. If used reflectively and repeatedly, historical empathy can also facilitate student contextualization of historical events into a broader understanding of the past and the development of a dispositional appreciation for the complexity of life and the situations faced by others across space and time. Engaging in historical empathy as a study of personal lived experiences helps us avoid the problematic "voiceless" government-centric approach, thereby leading to a richer, diverse understanding of such figures within a national and global context. Employing historical empathy in combination with the inclusion of the voices and experiences of the groups described above would enable students to view the past as a pluralist endeavor, one in which U.S. history is a chapter in a global story of peoples who were empowered, affected, or disaffected. Achieving this lofty goal would help prepare our students to become better global citizens but would also necessitate a shift in curricular approach that reaches into the heart of how we think about history itself.

Heritage versus History Approach to U.S. History Curriculum

Proponents of the currently predominant national heritage or collective memory approach to K-12 history education argue, "Democratic citizenship and effective participation in the determination of public policy require citizens to share a collective memory, organized into historical knowledge and belief" (McNeill, 1985). Featured elements of the national narrative include progress toward achieving national goals (Foster, 2006); emphasis on ethnic success stories while downplaying ethnic struggles and conflicts (VanSledright, 2008); current histories of immigrant groups primarily within the context of their lives in the United States that are virtually devoid of reference to experiences in their birth nation (Olneck, 1989); national development (VanSledright, 2008, p. 113); and a quest for freedom (Wertsch & O'Conner, 1991). Under the national heritage approach students learn "highly selective, sentimental, sanitized versions of American history [that represent] a severely simplified vision of how we came to be the society we are now" (Kammen, 1989, p. 139), and leave U.S. history courses knowing about the experiences of Americans through narrative accounts, but not necessarily believing what they have been told (Wertsch, 2000).

As noted by Banks (2008): "A major problem facing nation-states throughout the world is how to recognize and legitimize difference and yet construct an overarching national identity that incorporates the voices, experiences, and hopes of the diverse groups that compose it" (p. 133). Recognizing the importance of moving beyond a monocultural and exclusive narrative, we also realize the national narrative not only is composed of diverse peoples but also occurred in the past (Ladson-Billings, 2003). A K–12 history curriculum that emphasizes the sharing of a "collective memory" typically perceives the past through the lens of the present, thus placing little emphasis on historical thinking in general and historical empathy in particular.

Therefore, in the interest of promoting a more balanced and believable version of the past we propose that the national heritage approach be shelved in favor of an inquiry approach to K–12 history education that emphasizes the use of historical thinking in general and historical empathy specifically to help students construct a personal perspective on history rather than learning an accepted version of the past. Such an approach "involves searching for connections among disparate events to identify some developmental trend, causal patterns, or argumentative structure" (Barton & Levstik, 2004, p. 69). Students would be asked to "grapple with such concepts as causation, learn how to interpret documents, do historical detective work, and sharpen their historical imaginations" (Nash, Crabtree, & Dunn, 2000, p. 138) and learn an "American history [that] reveals the blemishes, leaves rough edges intact, and eschews cosmetics" (VanSledright, 2008, p. 121). An inquiry approach to history education enables students to take an active and critical role in "constructing an overarching national identity" as they explore the culture, experiences, and thinking of those who contributed to the identity within the context of those people's time. Ultimately, an inquiry approach to the U.S. history curriculum eschews dependence upon a single narrative, promotes historical thinking skills, emphasizes the use of historical evidence, and prepares students for life in a democratic society (Barton & Levstik, 2004).

Historical empathy would be central to this approach due to its emphasis on interpretation of historical evidence and the construction of historical understanding through students' personal relationships to the lived experiences of others. Yet it is easy to see why such an approach has not been more widely accepted in our schools. Historical thinking is not easy, and critics argue that expecting so much of students is why an inquiry approach fails to translate into practice (Leming, Ellington, & Porter-Magee, 2003). Yet, if we are to realize the truly humanizing potential that history offers, then we must understand that "historical thinking goes against the grain of how we ordinarily think, one of the reasons why it is much easier to learn names, dates, and stories than it is to change the basic mental structures we use to grasp the meaning of the past" (Wineburg, 2001, p. 7). Historical empathy, the most humanizing of historical thinking's various elements, turns up the proverbial volume of voices from the past and helps us understand all manner of lived experiences, which places it squarely "at the heart of historical inquiry" (Foster, 2001, p. 175).

Historical Empathy and the Narrative of Diverse Peoples, Not Nations

Here we provide four ways an inquiry approach that emphasizes historical empathy might globalize the chorus of voices in the K–12 history curriculum. They include the following:

1. Broaden and deepen the range of voices recently added to the curriculum.
2. Incorporate not simply the experiences of immigrants upon their arrival, but also capture their birth culture and the cultural interaction that occurred after their arrival.
3. Recognize that events in the United States occur contemporaneously with those worldwide. This requires using appropriate and representative voices to place such events into a global context.
4. Explore the narrative of other nations, particularly in relation to significant people and events in U.S. history, such as those found in other national history textbooks.

Broaden and Deepen the Range of Voices Recently Added to the Curriculum

Ironically, the press of instructional time causes educators to compress historical time, thus reducing curricular space, which makes simply including more voices challenging. With so much emphasis on the role of government in history, only a select few individuals have risen to prominence in the national narrative, resulting in disproportional individual representation. While Martin Luther King Jr., for example, unquestionably is a figure for the ages, how can students understand and appreciate him without learning about those who helped to pave the way for him? In broadening and deepening the chorus of African American voices, how about drawing upon Representative George H. White, the last African American elected to Congress until 1929, and his eloquent farewell address to Congress delivered January 29, 1901?

You may tie us then taunt us for a lack of bravery, but one day we will break the bonds. . . . You may withhold even the knowledge of how to read God's word and learn the way from earth to glory . . . but we remind you that there is plenty of room at the top, and we are climbing!

Mr. Chairman . . . I want to submit a brief recipe for the solution of the so-called "American Negro problem." He asks no special favors, but simply demands that he be given the same chance[s] for existence . . . that are accorded to kindred nationalities.

This, Mr. Chairman, is perhaps the Negroes' temporary farewell to . . . Congress; but let me say, phoenix-like, he will rise up some day and come again. These parting words are in behalf of an outraged, heartbroken, bruised, and bleeding, but God-fearing people, faithful, industrious, loyal people-rising people, full of potential force.

Historical empathy helps us to understand the deeper meaning behind Representative White's words and appreciate their prophetic nature. White, who was elected in the aftermath of Reconstruction in 1897, was able to attain a seat in Congress from the southern state of North Carolina at a time in which most of the Reconstruction reforms were being cast aside as Southerners regained control of their legislatures. In White's speech we see pride for his people, outrage at their treatment, and hope for the future. We can understand that the context of the time period has driven his understanding that he will most likely be the last African American in Congress for quite some time. He knows that the hope for African Americans lies within their own abilities, not in the freedoms granted to them by others. We can ascertain all of this by reading about George White's personal background, his early life, his education, and his time in Congress. By simply becoming more familiar with White's lived experiences and perspectives, we can understand why he was sad, outraged, and yet hopeful all at the same time. Including similar words by those such as Mary Church Terrell as she addressed the United Women's Club of Washington, D.C., on October 1, 1906, and Monroe Trotter in his *Guardian* newspaper would enable young people to realize that African Americans at the beginning of the twentieth century, and like King who followed, were represented by more than just soloists such as Booker T. Washington and W. E. B. Du Bois. This is but one example of broader inclusion, and we must remain mindful to avoid the problem of shifting relevance described earlier in this chapter by emphasizing the lived experiences of both men and women from different races, ethnicities, religions, and backgrounds during all historical eras, not just those in which interaction with the U.S. government was at play.

Capture Immigrants' Birth Culture and the Cultural Interaction That Occurred After Their Arrival

Beginning in the 1880s, the United States experienced relatively large-scale immigration first from China and then from Japan. In 1886 Hawaii and Japan signed a labor convention that led to a large number of Japanese first immigrating to Hawaii and then to the West Coast. While the Japanese immigrants' experience in Hawaii illustrated the possibilities of intercultural exchange as the Japanese Hawaiians eventually came to play a majority role, their experience on the U.S. mainland's West Coast proved quite different. During the first quarter of the twentieth century about 100,000 Japanese immigrated to the mainland, largely succeeding in agricultural businesses. A backlash occurred, resulting in the formation of the Japanese Exclusion League in 1905 and a diplomatic crisis between Japan and the United States caused by the San Francisco school board's decision to segregate Japanese and Japanese American students.

Historical empathy helps us to understand the thoughts, feelings, and motivations behind the groups involved in these events. Japanese immigrants to the United States followed a long line of immigrants from other nations (many of whom were discriminated against as well), and came to the United States in search of opportunity. These are motivations that we can easily appreciate and the rapid expansion of industry and agriculture on the West Coast made for a particularly attractive opportunity. However, many in the United States were opposed to the influx of Japanese workers because, as Theodore Roosevelt wrote to a friend, the Japanese "frugality, abstemiousness and clannishness make them formidable to our laboring class." (Roosevelt, 1905). As president of the United States, Roosevelt was opposed to the displacement of American workers. The San Francisco school board responded by declaring, "Our children should not be placed in any position where their youthful impressions may be affected by association with pupils of the Mongolian race" (Asian Society, n.d.). Roosevelt, while opposed to the immigration of Japan's working class, was angered by the school board's "wicked absurdity" and the possibility that relations with Japan could be endangered by the "crime against a friendly nation" (Roosevelt, 1906).

On the other side of the Pacific, Japanese pride had been damaged, and the responses ranged from that expressed in Tokyo newspaper by *Mainchi Shimbun*, "Stand up, Japanese nation! Our countrymen have been HUMILIATED on the other side of the Pacific" (As cited in Bailey, 1934) to that of Foreign Minister Hayashi (1906) who wrote that "the hostile demonstration in San Francisco has produced . . . a feeling of profound disappointment and sorrow [but] that feeling . . . is unaccompanied with any suggestion of retaliation. . . ." However, while angered by the public affront in full global view, the Japanese also had a strict policy against foreign immigration themselves and Roosevelt knew that "where they draw one kind of sharp line against us they have no right whatever to object to our drawing another kind of line against them." The end result was a gentleman's agreement that would rescind the segregation of Japanese students in San Francisco but would also prohibit future immigration for Japanese laborers.

With the help of historical empathy we can examine this incident from the viewpoints of the various parties involved, and in doing so we can come to understand how such an international incident can hold one meaning for one group of people yet can mean something else entirely to another. Not surprisingly, we see the battle over education as a recurring theme in U.S. history, not only for immigrants, but also for religious groups and politicians (e.g., the Scopes Trial). If the people are the "safe depositary of the ultimate powers of the society" and society should "inform their discretion by education" (Jefferson, 1820), then what better way to both personalize history for students and illustrate people's struggle to have their voices heard than by examining such groups' quest for an education and the motivations behind the actions of others who attempt to influence or even withhold it? Historical empathy has been demonstrated to help students understand not only the actions and decisions of traditionally significant figures such as Thomas Jefferson (Endacott, 2010) and Harry Truman (Doppen, 2000) but also the lived experiences of those with considerably less historical agency (Kohlmeier, 2006; Skolnick, Dulberg, & Maestra 2004). Again, this is but one example of how historical empathy can provide a glimpse into the experiences of Americans who quite naturally retained cultural and political ties to their former homelands. Wherever there is an immigrant influx both large (e.g., Irish Potato Famine) or small (e.g., Marshall Islanders), there are stories to be told about their experiences before, during, and after they came to the United States.

Recognize That Events in the United States Occur Contemporaneously with Those Worldwide

A third way to globalize U.S. history is situating it within the larger global context, which not only broadens the global chorus, but also moves students beyond an American exceptionalism approach, captured well by John O'Sullivan (1845, p. 5):

Hearing a Chorus of Voices

> It is now time for the opposition to the Annexation of Texas to cease . . . for the common duty of Patriotism to the Country to succeed. [Such opposition is] limiting our greatness and checking the fulfillment of our manifest destiny to overspread the continent allotted by Providence for the free development of our yearly multiplying millions.

Imagine if while learning about westward expansion and relations between Mexico and the United States prior to the Mexican-American War, students were presented with French consul Guerolt's observation in August 1845: "Instead of attacking California the Americans have peopled it; they have colonized it; they are taking it quietly bit by bit, while Mexico peacefully allows this province to be seized" (as cited in Anderson & Cayton, 2005, pp. 264-265). Now support Guerolt's observations with those of José Maria Tornel y Mendívil, 1837:

> From the state of Maine to Louisiana a call has been made in the public squares to recruit volunteers for the ranks of the rebels in Texas.
>
> [Mexican] character, our customs, our very rights have been painted in the darkest hues, while the crimes of the Texans have been applauded in the house of the President, in the halls of the capitol, in the marts of trade, in public meetings, in small towns, and even in the fields. The President of the Mexican republic was publicly executed in effigy in Philadelphia in an insulting and shameful burlesque.
>
> The Anglo-Americans, not content with having supplied the rebels with battleships to prey upon our commerce, . . . have protected them with their fleet and have captured ships of the Mexican squadron. . . .
>
> The loss of Texas will inevitably result in the loss of New Mexico and the Californias. Little by little our territory will be absorbed, until only an insignificant part is left to us. Our destiny will be similar to the sad lot of Poland. (as cited in Mintz, 2009, pp. 70-71)

Tornel Mendívil's words are particularly poignant in representing a Mexican perspective and in drawing connections to other similar historical events, such as how the Congressional War Hawks in 1812 used British seizure of U.S. ships as a pretext to declare war or how Chamberlain sought to appease Hitler by permitting him to annex the Sudetenland. Under the typical approach to American history, other nations from around the world make meaningful appearances only when their histories come into direct contact with the United States. As a result, conflicts and movements are viewed as America-centric even if they are also occurring elsewhere in the world simultaneously, thereby leaving our students with the perception of worldwide progress only as it relates to the United States, or of the United States as the sole driving force for this progress.

For example, in the three decades prior to the Nineteenth Amendment to the U.S. Constitution that granted women the right to vote, there were many countries in Europe as well as parts of Australia and Canada that had already granted suffrage. In the forty years that followed there would be many more nations in Europe, South America, and Asia that would also grant this basic right of citizenship. What were the experiences like for these women? In the United States we learn about important figures such as Elizabeth Cady Stanton or Susan B. Anthony, who were powerful agents of social change. Did the right to vote come about the same way for all women around the world? What about those who are still without political voice? How does the right of suffrage in the United States compare with the rest of the world across time? The use of historical empathy with figures such as Stanton and Anthony opens a window into the lives of iconic social agents, but failing to include the historical experiences of others from around the world who often struggled with the same issues at the same time only reinforces the notion of American exceptionalism. Other examples might include the civil rights movement in the United States in relation to the rise of apartheid in South Africa, or the similarities and differences between Manifest Destiny and other government-sponsored expansionist movements into native lands from around the world.

Explore the Narratives of Other Nations

Since history "textbooks serve as the arbiters of historical questions" (Wineburg, 1991, p. 84) for students, they represent the most powerful voice in the history curriculum and for the national narrative. While textbooks are authoritative, they are not monolithic. Drawing upon history textbooks of other nations, as well as of the United States, offers a final means to place U.S. history within a global context. Consider these dramatically different portrayals of the dispute between Texas and Mexico as found in a U.S. history textbook and a world history textbook published in the United States and a Mexican history textbook published in Mexico.

Were supporters of Texas, for example, "independence-minded settlers," "an alliance of Mexican liberals and American settlers," or "[separatists] . . . financed by the government of the USA"? Also, consider how both textbooks published in the United States mention slavery, which is absent from the Mexican textbook. Does this speak to the importance of slavery in the U.S. national narrative, while the Mexican textbook portrays Texas as an example of U.S. colonization? While there are no ready answers, such questions speak to the power of an inquiry approach as well as to the need to consider national textbooks as but another voice in globalizing the U.S. history chorus.

Figure 1: Mexican American War Textbook Excerpts from the United States and Mexico

America: Pathways to the Present (2005)
As their numbers swelled, these Americans [in Texas] demanded more political control. They wanted slavery to be guaranteed under Mexican law. When General Antonio Lopez de Santa Anna declared himself dictator of Mexico and stripped Texas of its rights of self government, Texans became united in the cause of independence. In October 1835, these independence-minded settlers clashed with Mexican troops, beginning the Texas War for Independence.
The Earth and Its Peoples: A Global History (2005)
Mexico also faced a grave threat from the United States. In the 1820's Mexico had encouraged Americans to immigrate to Texas, which at the time was part of Mexico. By the early 1830's Americans outnumbered Mexican nationals in Texas by four to one and were aggressively challenging Mexican laws such as the prohibition of slavery. In 1835 political turmoil in Mexico led to a rebellion in Texas by an alliance of Mexican liberals and American settlers. Mexico was defeated in a brief war, and in 1836 Texas gained its independence. (p. 606)
Historia 3 [Mexico] (2000)
There were many causes for the separation of Texas from the Mexican Republic. Among the most important the following stand out: • The expansionist policy of the USA. . . . • The erroneous politics of the Mexican government that permitted the uncontrollable colonization of Texas until the US population outnumbered the local Mexican population. • The separatist tendencies of Texas favored and financed by the government of the United States. (As cited in Lindaman and Ward, 2004, pp. 72-73.)

When students engage in historical empathy they are asked to consider the perspectives of those who lived in the past, but in doing so they also realize that their own perspectives are inevitably part of that process (VanSledright, 2001). In exploring the narratives of other nations, teachers have the opportunity to show their students how other people from different parts of the world read about the same historical events using a different lens, one that is heavily dependent upon the way in which history is portrayed in their own society. These alternate, or even competing, accounts of the past are part of somebody else's national narrative, but when considered in concert, they can become pieces of a global history that represents many national interests. It would be unwise to seek current news only from a single source and the same could be said about historical events that involved international or global interests. Exploring the narratives of other nations to consider another point of view does not guarantee that students will eschew their own perspective and come to see the past in the same way as somebody from Mexico, or Iraq, or Japan, but it is reasonable to hope that they will view these global and often competing perspectives as just as valid as their own.

Conclusion

Globalizing the U.S. history curriculum is a key component to preparing adolescents to become "citizens of the global community" and "make decisions to help solve the world's difficult problems" (Banks, 2008). Broadening the range of voices included in the curriculum is but the first step. Our students would also benefit from incorporating the birth culture and cultural interactions of immigrants to the United States, studying events in U.S. history contemporaneously with the rest of the world, and exploring the narrative of other nations. If students are to "realize history's humanizing qualities fully" (Wineburg, 2001, p. 6), then an inquiry approach to history that promotes historical empathy is critical to foster a more authentic interpretation of the past that promotes the contextualization of historical events into a broader understanding of global history as well as a dispositional appreciation for the lived experiences of others, both past and present. With these potential rewards, how can we not enable students to bathe in the richness of the global chorus of voices found in U.S. history?

References

Anderson, F., & Cayton, A. (2005). *The dominion of war: Empire and liberty in North America 1500–2000*. New York: Viking Press.

Asian Society. (n.d.). *Asian Americans then and now*. Retrieved from http://asiasociety.org/countries/traditions/asian-americans-then-and-now

Bailey, T.A. (1934). *Theodore Roosevelt and the Japanese American Crises*. Stanford University press.

Banks, J. A. (2008). Diversity, group identity, and citizenship education in a global age. *Educational Researcher*, 37(3), 129–139.

Barton, K. C. (2009). The denial of desire: How to make history education meaningless. In L. Symcox & A. Wilschut (Eds.), *National history standards: The problem of the canon and the future of teaching history* (pp. 265–282). Charlotte, NC: Information Age.

Barton, K., & Levstik, L. (2004). *Teaching history for the common good*. Mahwah, NJ: Erlbaum.

Bulliet, R. W., Crossley, P. K., Headrick, D. R., Hirsch, S. W., Johnson, L. L., & Northrup, D. (2005). *The Earth and its peoples: A global history* (3rd ed.). Boston: Houghton Mifflin.

Cayton, A., Perry, E. I., Reed, L., & Winkler, A. M. (2005). *America: Pathways to the present*. Needham, MA: Pearson.

Doppen, F. (2000). Teaching and learning multiple perspectives: The atomic bomb. *The Social Studies*, 91, 159–169.

Endacott, J. L. (2010). Reconsidering affective engagement in historical empathy. *Theory and Research in Social Education*, 38(1), 6–49.

FitzGerald, F. (1979). *America revised: History schoolbooks in the twentieth century.* Boston: Little, Brown.

Foster, S. (2001). Historical empathy in theory and practice: Some final thoughts. In O. Davis, E. Yeager, & S. Foster (Eds.), *Historical empathy and perspective taking in the social studies* (pp. 167-182). Mahwah, NJ: Rowman & Littlefield.

Foster, S. (2006). Whose history? Portrayal of immigrant groups in U.S. history textbooks, 1800-present. In S. J. Foster & K. A. Crawford (Eds.), *What shall we tell the children? International perspectives on school history textbooks* (pp. 155-178). Greenwich, CT: Information Age.

Hayashi, T. (1906, October 23). *Telegram from Hayashi, Japanese Foreign Minister, to K. Uyeno, the Japanese Consul at San Francisco.* Retrieved from http:// bss.sfsu.edu/ waldrep/hist642/ telegrams.html

Jefferson, T. (1820). Letter to William C. Jarvis. In Bergh, A. (Ed.) *The Writings of Thomas Jefferson*, 20 Vols., 15:278 Washington, D.C., 1903-04. Retrieved from http:// www.constitution.org/tj/jeff.htm

Kammen, M. (1989). History is our heritage: The past in contemporary American culture. In P. Gagnon (Ed.), *Historical literacy: The case for history in American education* (pp. 138-156). Boston: Houghton Mifflin.

King, J. E. (1995). Culture-centered knowledge: Black studies, curriculum transformation, and social action. In J. A. Banks & C. M. Banks (Eds.), *Handbook of research on multicultural education* (pp. 265-290). New York: Macmillan.

Kohlmeier, J. (2006). "Couldn't she just leave?": The relationship between consistently using class discussions and the development of historical empathy in a 9th grade world history course. *Theory and Research in Social Education,* 34(1), 34-57.

Ladson-Billings, G. (2003). *Critical race theory perspectives on the social studies: The profession, policies, and curriculum.* Greenwich, CT: Information Age.

Leming, J., Ellington, L., & Porter-Magee, K. (2003). *Where did social studies go wrong?* Washington, DC: Fordham Foundation. Retrieved from http:// www.edexcellence.net/doc/ ContrariansFull.pdf.

Lerner, R., Nagai, A. K., & Rothman, S. (1995). *Molding the good citizen: The politics of high school history texts.* Westport, CT: Praeger.

Lindaman, D., & Ward, K. (2004). *History lessons: How textbooks around the world portray U.S. history.* New York: New Press.

Lopez, J. de J. N., et al. (2000). *Historia 3.* Mexico: Santillana. [As found in Lindaman, D., & Ward, K. (2004). *History lessons: How textbooks around the world portray U.S. history.* New York: New Press.]

McNeill, W. H. (1985). *Why study history?* American Historical Association. Retrieved from http://www.historians.org/pubs/archives/whmcneillwhystudyhistory.htm

Nash, G. B., Crabtree, C., & Dunn, R. E. (2000). *History on trial: Culture wars and the teaching of the past.* New York: Vintage Books.

Olneck, M. R. (1989). Americanization and the education of immigrants, 1900-1925: An analysis of symbolic action. *American Journal of Education,* 92, 398-423.

O'Sullivan, J. (1845). Annexation. *United States Magazine and Democratic Review,* 17(1), 5-10.

Paxton, R. J. (1999). A deafening silence: History textbooks and the students who read them. *Review of Educational Research,* 69(3), 315-339.

Paxton, R. J. (2002). The influence of author visibility on high school students solving a historical problem. *Cognition and Instruction,* 20(2), 197-248.

Portal, C. (1990). Empathy. *Teaching History,* 58, 36-38.

Roosevelt, T. Letter, May 6, 1905, to George Kennan, *The Selected Letters of Theodore Roosevelt*, edited by H.W. Brands. New York: Cooper Square Press, 2001, p. 380.

Roosevelt. T. *"Annual Message to Congress."* 4 Dec. 1906.

Skolnick, J., Dulberg, N., & Maestra, T. (2004). *Through other eyes: Developing empathy and multicultural perspectives in the social studies* (2nd ed.). Toronto, Ontario, Canada: Pippin.

Tornel y Mendivíl, J. M. (1837). Relations between Texas, the United States of America and the Mexican Republic. In Mintz, S. (Ed.) (2009). *Mexican-American voices: A documentary reader*. (2nd ed.). West Sussex: John Wiley & Sons.

VanSledright, B. (2001). From empathetic regard to self-understanding: Im/positionality, empathy, and historical contextualization. In O. Davis, E. Yeager, & S. Foster (Eds.), *Historical empathy and perspective taking in the social studies* (pp. 51–68). Lanham, MD: Rowman & Littlefield.

VanSledright, B. (2008, February). Narratives of nation-state, historical knowledge, and school history education. *Review of Research in Education*, 32(1), 109–146.

Wertsch, J. V. (2000). Is it possible to teach beliefs, as well as knowledge about history? In P. Stearns, P. Seixas, & S. Wineburg (Eds.), *Cognitive and instructional processes in history and the social sciences* (pp. 38–50). New York: New York University Press.

Wertsch, J. V., & O'Connor, K. (1991). Multi-voicedness in historical representation: American college students' accounts of the origin of the US. *Journal of Narrative and Life History*, 4, 295–310.

White, G. H. (1901). *Farewell address to Congress, U.S. House of Representatives, January 29, 1901*. Retrieved from: http://www.edchange.org/multicultural/speeches/george_white_farewell.html

Wineburg, S. S. (1991). On the reading of historical texts: Notes on the breach between school and academy. *American Educational Research Journal*, 28, 495–519.

Wineburg, S. (2001). *Historical thinking and other unnatural acts*. Philadelphia: Temple University Press.

CHAPTER FOURTEEN

Global Education for Critical Geography

Jason R. Harshman

The Orient was almost a European invention

Edward Said, *Orientalism*, 1978

To better understand the processes that shape how the world is organized and imagined, the study of geography must incorporate more than the extent to which students are able to identify cities and mountains on a map. Geography, John Dewey (1897) argued, is more than the classification of facts, it is the way an "individual feels and thinks the world" (p. 168). Integrating critical geography into global education involves engaging students in a deeper appreciation for the diversities that make up human and cultural geography, while also questioning the local and global processes that shape everyday experiences of space and power (Gruenewald, 2003; Helfenbein, 2006; Massey, 2007). If students are to decolonize and decenter their worldview, educators need to adopt a more globally minded approach to teaching and learning world geography (Merryfield, 2001).

This chapter begins with a brief overview of how space has been theorized and defined in the fields of geography and education to illustrate the relationship between critical geography and global education. The next section addresses the construction of place and the extent to which technologies and globalization complicate teaching about here and there. The concepts advanced in this chapter are then applied to the study of two countries, Turkey and China, to illustrate how critical approaches to geography and global education can come together to foster new ways of thinking about the world.

Connecting Critical Geography and Global Education

Critical geography "is concerned with examining how spatial relationships shape culture, identity, and social relationships" (Gruenewald, 2003, p. 628). Likewise, meanings inscribed in a space are constructed through a network of social institutions and lived experiences (Soja, 1989). While scholars in geography and other disciplines have critically examined the relationships between power, space, time, and cultural production (Foucault, 1986; Harvey, 1996; Lefebvre, 1974; Massey, 1993; Soja, 1989, 1996), space remains under examined in education (Gruenewald, 2003; Gulson & Symes, 2007). Developing a student's global sense of place involves interrogating the construction of boundaries that serve to reinforce thinking about "us" and "them" and is best accomplished through global education (Harvey, 1996; Helfenbein, 2006; Massey, 1993; Said, 1978).

Since Hanvey's (1975) seminal work, there have been multiple conceptualizations of what global education encompasses (Case, 1993; Kirkwood, 2001; Merryfield, 2001; Subedi, 2010). Across these conceptualizations, interconnectedness, perspective consciousness, open-mindedness, and a sense that global education goes beyond knowledge to include attitudinal development and action stand as commonly agreed-upon components (Case, 1993; Merryfield, 1991; Pike, 2000).

According to the National Council for the Social Studies (NCSS) position statement on preparing students for a global community, "a global perspective is attentive to the interconnectedness of the human and natural environment" and "in studying the traditions, history, and current challenges of other cultures, the perspective consciousness of our students must be raised and ethnocentric barriers must be addressed" (NCSS, 2001). The barriers a globally minded educator looks to teach beyond include how nation-states have come to be conceptualized and how students imagine the local and the global organization of places and people. The goals and values of global education and the development of democratically minded global citizens cannot be separated from critical geography (Gaudelli & Heilman, 2009).

Deterritorializing Space and Culture

The set of spatial structures through which people order their knowledge of the world according to continents, countries, and cities is known as "metageography" (Lewis & Wigen, 1997, p. 8). Metageography is a point of analysis used to better understand the power of ideology related to using geography as an organizing principle. This manner of thinking about places as "other" is similarly adopted within a location as political institutions seek to create uniformity despite ethnically, racially, linguistically, and religiously diverse populations within a given nation (p. 9). We understand this historical formation to be the construction of the nation-state.

Although a history of hybridity and transculturation exists in most regions of the world, the processes and institutions responsible for constructing a coherent nation must exert a level of violence to unite people around a shared identity (Mitchell, 2002). Maintaining perceivably common bonds between citizens involves building nationalism through national symbols, stories, and

Global Education for Critical Geography

customs to unite otherwise different peoples. The unification of peoples in a shared place who may not ever meet through the use of symbols and images contributes to the formation of an "imagined community" (Anderson, 1991). Benedict Anderson's focus on the role of print capitalism in orchestrating a sense of shared space at a national level has morphed into an electronic form that transcends national boundaries due to advancements in consumer technologies (i.e., portable personal computers and smart phones), thus displacing how people think about distance and place. These global processes that have created more interconnected, mobile populations are what Arjun Appadurai (1996) calls "scapes," and they have served to disrupt national boundaries and deterritorialize culture.

"Mediascapes" produce and disseminate information electronically and have become more accessible to a growing audience of private and public users around the world (Appadurai, 1996, p. 35). Mediascapes, however, can be manipulated to highlight differences as part of a national and transnational project undertaken by nation-states and media producers to mold ideologies for political purposes (p. 15). Politicizing media images to support state ideologies against counter-ideologies that aim to undermine state power are "ideoscapes" (p. 36). Important to understanding the function of ideoscapes is that they are flexible, and often the meaning ascribed to an image, idea, or place differs from how they are received. Increased accessibility and mobility due to "technoscapes" have deterritorialized culture through the creation of new markets for film, music, and cultural commodities in multiple places (p. 38). A "technoscape" is a fluid, global configuration that includes high and low, mechanical and informational technologies that move across multiple, previously impenetrable boundaries (p. 34). Examining the connections between these "scapes" and curricula is important if students are to understand changes in the global society and develop a more critical understanding of what shapes place, culture, and identity.

Within the scapes that carry images and cultures from place to place are "codes" that serve to reify the meanings that institutions want to convey about peoples, places, events, and ideas (Hall, 2006). Meaning making through media and education is an ongoing process of encoding and decoding messages that are widely distributed because the codes are often learned at an early age (racism, sexism, and "othering" in general) and thus reach near-universal or naturalized meaning. Yet, as global communication networks expand, culture continues to be deterritorialized, and the movements of people continue to change, diversify, and pluralize the cultural identity of the nation-state (Hall, 1999). Consequently, what is meant by "here" and "there" is continually changing, meaning that places encompass multiple identities.

To take a critical approach to teaching and learning social studies and history is to render the histories that have been long presented as complete and factual problematic (Segall, 1999). In terms of critical geography, one must interrogate the ways in which meaning has been ascribed to spaces through hierarchies of power in order to construct an identity to and for a place (Helfenbein, 2006). Taken together, these theories and perspectives on how we imagine space, identity, and power at local, national, and international levels, in a time of hyperglobal interconnectedness and movement, exemplify the need for critical geography. For educators, the question remains: Since the natural world does not conform to arbitrary political boundaries, why should peoples and cultures?

Developing a Globally Minded Critical Geography Education

The realities of our mobile, global society require that students in the United States not only think about and understand diversity "here" and "there," but that they also possess the knowledge and skills necessary for cross-cultural interactions (Merryfield & Wilson, 2005; Nganga, 2009). As schools and classrooms continue to become more richly diverse, adopting a critical, global education-based approach to studying geography is necessary so that students recognize and understand their place in the network of space, power, culture, and identity.

This section provides ideas on how to take a more critical approach to teaching about the physical, cultural, and human geography of Turkey and China. Focus is given to Turkey and China as both nations contain a diverse population, a varied physical landscape, and multiple cultural practices, and each plays an increasingly influential role in the global economy. The section about Turkey examines a complex question: Where is Turkey? The essential question for the lesson ideas for teaching about China is, Who lives in China?

Where Is Turkey?

In the spring of 2011, I had the good fortune to travel to Turkey as part of a study tour to learn from educators and students about their country and how they conceptualize global-mindedness. My travel involved frequent trips across the Bosporus, a strait in Turkey that separates the continents of Europe and Asia. Each passage across the bridges meant I was moving from one continent to another without leaving the city of Istanbul, let alone the country of Turkey.

Global Education for Critical Geography 173

Considering Turkey's geographic location, political and economic relations in the region and world, and cultural and religious ties to Asia, Europe, and the Middle East, the question Where is Turkey? must be asked.

In keeping with the tenets of global education, it is important to listen to the people in the places we intend to study. Students need to "analyze global problems and issues that may be caused because of economic and political connections among countries" (Açikalin, 2010, p. 255). Since Turkey stands at a physical crossroads (the intersection of what we understand as two different continents) as well as a political, cultural, and economical crossroads (e.g., Is membership in the European Union [EU] in Turkey's best interest? How does the Turkish government balance secular practices in a Muslim majority country?), Turkey serves as an important case study for understanding global interconnectedness and critical geography.

At the outset of a unit of study about Turkey, teachers should ask students what they know or think they know about Turkey, such as where the country is located, who lives there, and whether they think of Turkey as a modern country or not (this is also a good opportunity to complicate the tendency to conflate the meaning of "modern" with Western or American notions). Next, present students with a map that includes Turkey and the surrounding regions of Europe, the Middle East, and Asia and ask the question Where is Turkey? Possible answers should include (A) Europe; (B) Asia; (C) the Middle East; and (D) All of the above.[1]

After students have had an opportunity to select an answer, ask them why they chose that region in which to locate Turkey. Following a conversation that may include generalizations and stereotypes (Merryfield & Wilson, 2005), the teacher should present students with three more maps: one of Europe, one of the Middle East, and one of Asia, each with Turkey included. The following question can then be posed: How does where Turkey is located on a map affect your earlier answer? Working through the images of where the nation-state of Turkey is physically located opens the door for conversations about the complexities of human and cultural geography. Building upon the principles of global education, this part of the lesson should emphasize fluidity, multiplicity, movement, and flexibility when discussing borders, cultures, and space.

It is recommended that the teacher use images from around the country: rural and urban; men, women, and children; industrial and agricultural locations; and several photos of Turkey, from the beaches of southern Turkey to the arid lands of eastern Turkey. Be sure to include both historic and cultural sites, but also ask how Turkey is affected by and contributes to globalization. Discussion questions for teaching about Turkey include the following: Who is Turkish? What does it mean to be Turkish? How are conceptualizations and legal statutes regarding citizenship affected by globalization? By avoiding what may be perceived as the exotic elements of Turkey, teachers will help students develop a more complex understanding of how nation-states and the arbitrariness of their borders are constructed, along with the diverse, hybrid cultural practices that exist in Turkey (Akinoğlu, 2004). While these ideas are presented as part of a critical examination of Turkey's cultural, physical, and human geography, such issues are not unique to Turkey, as is addressed in the next section on human geography in China.

Who Lives in China?

Conflating culture with place, race, and ethnicity can serve as a political ploy to deny the reality of global systems. Examining history and how we think of the contemporary world in a postmodern sense allows for a challenging of what had been posited by modernists as objective truth. Tambiah (2000) disrupts the Western hegemonic model of history and geography by arguing for a reading of world history from an Asian perspective, with attention to the movements of people, goods, and ideas that have traveled throughout the area for centuries.

According to the U.S. Central Intelligence Agency's (2012) entry on China in *The World Factbook*, as of 2012, China is a country of over 1.3 billion people. While the Han constitute the ethnic majority, there are fifty-six ethnic groups recognized in China. Furthermore, although officially an atheistic country, recognized religions include Daoism, Buddhism, Christianity, and Islam. Learning about the diversity within China raises important questions to explore with students when discussing critical human and cultural geography: What do we think when we hear or say "China"? How do members of minority groups in China think about citizenship, belonging, and/or identity? What roles do nationalism and globalization play in relation to what it means to be a Muslim or Hui in China?

From the deserts and mountains in the north and west, to the mixture of rugged terrain and plateaus of the south, to the rivers and coastal areas of the east, China, the fourth largest country in the world in terms of total area, is geographically diverse. Consequently, the relationship between culture and location, along with ethnicity, religion, and economics, is very important in helping students develop a better understanding of who lives in China. The diversity within China, similar to that of Turkey and most other countries, is an opportunity for educators and students to investigate issues of power in defining national culture, and how majority and minority groups are not only constructed but the efforts taken to maintain them in a time of global migrations.

The proximity of western China to what many in the United States call the Middle East means that many Uyghurs, a Turkic-speaking ethnic group in Central and Eastern Asia, live within the boundary of China. The Uyghur diaspora, however, expands across multiple borders into multiple nation-states including Uzbekistan, Afghanistan, Kazakhstan, and Turkey, to name a few. As a transnational, marginalized population, Uyghurs face tremendous persecution, and while their struggle for space continues, such efforts stand in contrast to the laws of nation-statehood (Bovingdon, 2010).[2]

Global education includes teaching about controversial and complex issues, while also drawing upon primary sources to provide students with multiple perspectives for the issues and places they study (Case, 1993). To truly complicate discussions of power, ethnic majorities and minorities, and critical human geography, teaching about China must include more than the current political and economic presence of the Chinese government throughout the world. Working with a political map of China, teachers should begin at the borders and challenge students to avoid speaking about China as an ethnically homogeneous place. Following an analysis of who lives there and crosses the border, primary sources authored by members of multiple ethnic groups living in China should be used to develop students' perspective and consciousness about what it means to be Chinese when you are not a member of the ethnic majority. This approach to teaching critical human and cultural geography involves discussions about identity, resistance, citizenship, and the importance of social justice in relation to space (Soja, 2010).

Advancing Critical Geography in Global Education

Critical approaches to global education require that students not only examine the past and present but develop skills for "managing the complexities of the future" (Kirkwood, 2001). As people and ideas continue to cross borders, it is essential for students to understand that the establishment of and attempt to maintain a monocultural nation-state contribute to struggles regarding individual and national identities. Understanding that the borders, names, and categories by which the world is organized stem from a long history of imperialism and its ramifications constitutes a responsibility that global educators must be willing to take on (Willinsky, 1998). The implementation of such practices can include, but should not be limited to, the following: 1) a more critical approach to the study of cultural, human, and physical geography when referencing borders, rights of ethnic minorities, and diasporas; 2) inclusion of primary source materials, including images that counter stereotypes, in order to represent the diversity of regions rather than relying upon Western-constructed narratives of the world; 3) development of cross-cultural understanding and perspective consciousness necessary for developing global-mindedness; and 4) opportunities for students to critically examine and reflect upon their understanding of space, power, culture, time, and perspective in relation to globalization.

While there is no denying that the nation-state remains integral to the maintenance of an ideology of national allegiance prominent around the world, global mediascapes and technoscapes, along with waves of migration, have transformed how people imagine the world. As individuals exist at the intersection of multiple global forces that inform their perspective of the world and their identity formation, these same principles must be acknowledged and developed in our classrooms.

Notes

1. Go to http://www.worldatlas.com to find regional and continental maps of Europe, Asia, and the Middle East that each include Turkey.
2. One source educators can use to obtain a Uyghur perspective is http:// www.iuhrdf.org/ .

References

Açikalin, M. (2010). The influence of global education on the Turkish social studies curriculum. *The Social Studies*, 101, 254–259.
Akinoğlu, O. (2004). An analysis of 2004 Turkish social studies curriculum in the light of new millennium trends. *Social Behavior and Personality*, 36(6), 791–798.
Anderson, B. (1991). *Imagined communities: Reflections on the origin and spread of nationalism* (2nd ed.). London: Verso.
Appadurai, A. (1996). *Modernity at large: Cultural dimensions of globalization*. Minneapolis: University of Minnesota Press.
Bovingdon, G. (2010). *The Uyghurs: Strangers in their own land*. New York: Columbia University Press.
Case, R. (1993). Key elements of a global perspective. *Social Education*, 57(6), 318–325.
Central Intelligence Agency. (2012). *East and Southeast Asia: China. The world factbook*. Retrieved from https://www.cia.gov/library/publications/the-world-factbook/geos/ch.html
Dewey, J. (1897). The psychological aspect of the school curriculum. *Educational Review*, 13, 356–369.
Foucault, M. (1986). Of other spaces. *Diacritics*, 16(1), 22–27.
Gaudelli, W., & Heilman, E. (2009). Reconceptualizing geography as democratic global citizenship education. *Teachers College Record*, 111(11), 2647–2677.
Gruenewald, D. (2003). Foundations of place: A multidisciplinary framework for place conscious education. *American Educational Research Journal*, 40(3), 619–654.
Gulson, K., & Symes, C. (Eds.). (2007). *Spatial theories of education: Policy and geography matters*. New York: Routledge.
Hall, S. (1999). Thinking the diaspora: Home-thoughts from abroad. *Small Axe*, 6, 1–18.

Hall, S. (2006). Encoding/decoding. In M. Durham & D. Kellner (Eds.), *Media and cultural studies: Key works* (pp. 163-173). London: Blackwell.
Hanvey, R. (1975). *An attainable global perspective*. New York: Center for War/Peace Studies.
Harvey, D. (1996). *Justice, nature, and the geography of difference*. Malden, MA: Blackwell.
Helfenbein, R., Jr. (2006). Space, place, and identity in the teaching of history: Using critical geography to teach teachers in the American South. In A. Segall, E. Heilman, & C. Cherryholmes (Eds.), *Social studies—the next generation: Researching in the post-modern* (pp. 111-124). New York: Lang.
Kirkwood, T. (2001). Our global age requires global education: Clarifying definitional ambiguities. *The Social Studies, 92*, 1-16.
Lefebvre, H. (1974). *The production of space*. Cambridge, MA: Blackwell.
Lewis, M., & Wigen, K. (1997). *The myth of continents: A critique of metageography*. Berkeley: University of California Press.
Massey, D. (1993). Power-geometry and a progressive sense of place. In J. Bird, B. Curtis, G. Robertson, & L. Tickner (Eds.), *Mapping the futures: Local cultures, global change* (pp. 59-69). New York: Routledge.
Massey, D. (2007). *World city*. Cambridge: Polity.
Merryfield, M. M. (1991). Preparing American secondary social studies teachers to teach with a global perspective: A status report. *Journal of Teacher Education, 42*, 11-20.
Merryfield, M. M. (2001). Moving the center of global education: From imperial world views that divide the world to double consciousness, contrapuntal pedagogy, hybridity, and cross-cultural competence. In W. B. Stanley (Ed.), *Social studies: Research, priorities and prospects* (pp. 179-207). Greenwich, CT: Information Age.
Merryfield, M. M., & Wilson, A. (2005). *Social studies and the world: Teaching global perspectives*. Silver Spring, MD: National Council for the Social Studies.
Mintz, S. (2009). *Mexican American Voices: A documentary reader*. Malden, MA: Blackwell.
Mitchell, T. (2002). Heritage and violence. In T. Mitchell (Ed.), *Rule of experts: Egypt, technopolitics, modernity* (pp. 1-25). Ann Arbor: University of Michigan Press.
National Council for the Social Studies. (2001). *Preparing citizens for a global community*. Retrieved from http://www.ncss.org/positions/global
Nganga, L. (2009). Global and cultural education prepares preservice teachers to work in rural public schools. Teaching for social change in the 21st century. *Journal of Education Research, 3*(1), 149-160.
Pike, G. (2000). Global education and national identity: In pursuit of meaning. *Theory into Practice, 39*(2), 64-73.
Said, E. (1978). *Orientalism*. New York: Vintage Books.
Segall, A. (1999). Critical history: Implications for history and social studies education. *Theory and research in social education, 27*(3), 358-374.
Soja, E. (1989). *Postmodern geographies: The reassertion of space in critical social theory*. London: Verso.
Soja, E. (1996). *Thirdspace*. Cambridge, MA: Blackwell.
Soja, E. (2010). *Seeking spatial justice*. Minneapolis: University of Minnesota Press.
Subedi, B. (Ed.). (2010). *Critical global perspectives: Rethinking knowledge about global societies*. Charlotte, NC: Information Age.
Tambiah, S. (2000). Transnational movements, diaspora, and multiple modernities. *Daedalus, 129*(1), 163-194.
Willinsky, J. (1998). *Learning to divide the world: Education at empire's end*. Minneapolis: University of Minnesota Press.

CHAPTER FIFTEEN

Blogging for Global Literacy and Cross-cultural Awareness

Kenneth T. Carano
Daniel W. Stuckart

The globalization of economic, political, technological, and environmental systems has permanently altered the knowledge and skills that young people need to become effective citizens (Merryfield, 2000). The scale with which these systems are transforming many modern societies is unprecedented; therefore, globalization has the possibility of expanding the critical, imaginative, and ethical dimensions of education (Heilman, 2009). As a result, students need to understand these global interconnections if they are to take advantage of the opportunity to attain their full potential (Tye, 1999) and develop the skills enabling them to interact effectively with people different from themselves. One method for doing this is through digital technology, which has the potential of breaking the walls of the traditional classroom (Carano & Berson, 2007).

The invention of the Web browser has facilitated a historic change in global connectedness, and a significant consequence has been the online phenomenon of social networking sites (Kirkpatrick, 2006), which are Internet sites where people can come together to communicate with one another (Metz, 2006). Their popularity is underscored by the fact that the online social networking site Facebook has surpassed Google to become the top source for Internet traffic (Evangelista, 2010). Cognizant of this increasingly interconnected world and the profound potential of utilizing this technology in the social studies classroom, this study explores using Weblogs, or blogs, as a pedagogical tool in the development of global literacy and cross-cultural awareness in secondary school students. Students have used blogs as an educational tool to engage with academic content, including as a means to try to combat students' cultural stereotyping (Carano, Keefer, & Berson, 2008).

The study fits well with the position statement of the National Council for the Social Studies on global education, which supports the use of technology to develop global experiences and prepare students for global citizenship (NCSS, 2010). In addition, a major topic in the Advanced Placement Human Geography curriculum includes a study of the relationship between culture and physical landscapes (College Board, 2011). The framework also aligns with seminal social science research related to meaningful, cross-cultural awareness experiences and global literacy pedagogy (Hanvey, 1976; Merryfield & Wilson, 2005).

Review of the Literature
Global Awareness

The past decade has seen a renaissance of literature emphasizing the need for teaching for global awareness among K–12 students in order to prepare them for an interconnected world (see, e.g., Carano & Berson, 2007; Kirkwood-Tucker, 2009a, 2009b; Zong, 2009). Based upon the overview of the global awareness literature and the underlying components of its dominant pedagogy, four dimensions were identified (see Table 1) as necessary components to be applied in the classroom.

Table 1. Four Dimensions of a Global Awareness

Dimension	Definition
Cross-cultural awareness	Understanding the uniqueness of the individual and culture.
Global literacy	Skills and awareness needed to work in a globalized world.
Service-learning	Process of learning and developing through active participation in organized service experiences that meet community needs both locally and globally.
Social justice	Analyzing the issues of power structure arrangements and the fair distribution of advantages, assets, and benefits among all members of a society.

The cross-cultural awareness dimension, first articulated by Hanvey (1976), focuses on understanding the uniqueness of the individual and culture and includes developing empathy for others and recognizing that the dynamics of cultures are in a constant state of fluctuation (Imbert, 2004; Merryfield, 2001). The second dimension, global literacy, consists of five subcategories: state of the planet awareness, knowledge of global dynamics, awareness of human choices, media literacy, and research and thinking skills. Within the framework of state of the planet awareness, some of the issues that global educators identify as being pertinent include the awareness of different global belief systems, political systems, economic systems, and population issues (Merryfield &

Wilson, 2005). Knowledge of global dynamics is the awareness that world events are interconnected and have unanticipated consequences (Hanvey, 1976). In addition, it involves analyzing the shrinking of space and time and how it has fostered interconnections, raised awareness around the globe, and changed the nature of social space (Pike & Selby, 2000). Another concept identified by Hanvey (1976), awareness of human choices, is the ability to realize the long-range implications of choice from multiple perspectives. The fourth subcategory, media literacy, encompasses gaining critical technological skills that enable students to have an enhanced personal life and learn about the world (Merryfield & Wilson, 2005). Last, research and thinking skills include analytical thinking and problem-solving skills, which are increasingly important in an interconnected world if students eventually are to be able to compete economically (Joftus, 2004) and become globally competent citizens (Kirkwood, 2001). These capabilities also entail synthesizing skills along with the ability to detect bias and unstated assumptions so that students can acquire the decision-making skills that they need as citizens of a democratic society (Merryfield & Wilson, 2005).

Service-learning has two subcategories: community service and sustainable development. Participating in community service means being actively involved in improving the human condition at the local or global level (Kirkwood, 2001). Sustainable development refers to the present generation meeting its needs, while limiting its use of nonrenewable sources to meet the needs of future generations (Banks et al., 2005). The last global education dimension, social justice, has the following subcategories: legacy of colonialism, human rights, and power/influence of the media. Legacy of colonialism examines the origins and assumptions underlying a Eurocentric framework and analyzes alternative frameworks (Merryfield, 2001). Human rights encompass the fair treatment of all social groups (Landorf & Nevin, 2007). The concept of the power/influence of the media refers to the necessity of students' unlearning the often-exaggerated information promulgated because of media stereotypes (Cortes, 2000).

Global Awareness and Technology

Ferriter (2010) described one of the few appraisals of technology's influence on K–12 school-age students' global awareness when writing about the Flat Classroom Project, in which teachers in the United States and Bangladesh used digital tools to foster international collaboration between U.S. and Bangladeshi students. The students used online discussion boards, video conferences, instant messages, and e-mails to communicate. The investigators concluded that students built impressions based upon actual people rather than stereotypes and expressed a greater understanding of and appreciation for their international peers.

While the research focusing on technology's effect on the global awareness of K–12 students is limited, "teacher education scholars have been particularly interested in exploring the potential of Internet-based technology, such as computer-mediated communication, in building cross-cultural understanding and promoting global awareness among preservice teachers" (Zong, 2009, p. 80). Merryfield (2000) analyzed the incorporation of online threaded discussions in graduate courses in social studies and global education by comparing the topics, their depth, and patterns of interaction of Internet discussions with face-to-face discussions. She found that online class discussions tended to be more equitable in the distribution of student comments and contained an increase in the depth of content discussed. In addition, research indicated that by utilizing online discussion, cross-cultural learning improved.

Zong (2002) followed two pre-service teachers to examine the influence of participating in a computer-mediated international communication project that discussed world issues to see how the course influenced their understanding of global education. The analysis suggested that using the Internet to communicate in this manner has the potential of improving pre-service teachers' global literacy, gaining an appreciation of other people's perspectives, and is a motivator for teaching from global perspectives. Gaudelli (2006) studied the experiences of two beginning social studies teachers and found that an Internet distance learning course for global education made an impression on their motivation and understanding of teaching global education.

Objectives and Purpose of the Study

In spring 2011, the authors studied an Advanced Placement Human Geography classroom of students' blogging with residents of Malaysia and Swaziland. The study took place at a public high school located in the southeastern United States. The study commenced after the AP exam in May and explored the following research questions:

1. In what ways do students develop global literacy as the result of blogging with Swaziland and Malaysian residents?
2. To what extent is student cross-cultural awareness enhanced as a result of the blogging experience?

Methods

Research Framework. Two dimensions of global awareness were explored, global literacy and cross-cultural awareness. Global literacy refers to the ability to understand world conditions, trends, interconnections, unanticipated consequences, and specific skills associated with this understanding (Hanvey, 1976; Merryfield & Wilson, 2005). Based on the literature, five subcategories in which students should become proficient were identified within global literacy. The categories include the following:

1. State of the planet awareness
2. Knowledge of global dynamics
3. Awareness of human choices
4. Media literacy
5. Research and thinking skills

Cross-cultural awareness is the ability to perceive one's own culture from other vantage points, and being able to live in another culture as opposed to simply having the ability to live with it (Hanvey, 1976). The authors adapted Hanvey's four hierarchical levels as a framework when evaluating increased cross-cultural awareness, which resulted in the following four general categories:

1. Awareness of superficial or visible traits; stereotypes
2. Awareness of cultural traits that contrast markedly with one's held stereotypes when confronted with a culturally conflicting situation
3. An intellectual analysis of cultural traits that contrast with one's own
4. Awareness of how another culture feels from an insider's point of view

Research Setting. The research setting was a classroom in an ethnically diverse high school, located in the southeastern United States, populated with about 1,500 students. About 41 percent of students were eligible for free or reduced-price lunch. The school received the top grade in the state's grading system during the previous academic year. The teacher was a former Peace Corps volunteer, and had worked as a social studies teacher at the school for over eight years.

Research Participants. There were eighteen student participants. All students in the course returned the parent/guardian and student permission forms. The class consisted of three freshmen, one sophomore, five juniors, and nine seniors. The GPA range was 1.75 to 3.75. Five male and thirteen female students reported an ethnic breakdown of seven white, four black, four Hispanic, two mixed ethnicities, and one Asian. The activity was incorporated as a part of the final exam course grade, with students evaluated based on participation.

Analysis of Results. The data sources include pre- and post-questionnaires, blog postings, and interviews. While analyzing the qualitative data, the authors used the framework developed by Miles and Huberman (1994) to describe the major phases of data analysis: data reduction, data display, and conclusion drawing and verification. In order to establish an initial attitudinal baseline in global literacy and cross-cultural awareness, the researchers administered pre-questionnaires prior to beginning the exercise. Next, after reading articles and Web sites, watching videos, and discussing the geography and culture of each area, students completed a second questionnaire. Over the next two weeks, students blogged with the cultural experts (i.e., residents) living in Swaziland and Malaysia. At the end of the two weeks, students posted a final blog reflection on what they had learned citing evidence from earlier posts. Subsequently, all students completed open-ended blog experience questions for Questionnaires 3 and 10. (See Appendices). Nine students also participated in structured interviews.

Discussion

Qualitative evidence suggests that participants learned and perceived educational value and satisfaction by blogging with residents in Swaziland and Malaysia. Many themes emanating from the data, such as the breaking of stereotypes and the underlying influence of religion, focused on specific elements that demonstrated an increased cross-cultural awareness. Further, evidence suggests the development of global literacy, specifically the subthemes of the awareness of the state of the planet and knowledge of global dynamics. Participants also repeatedly stated an enjoyment in developing a personal bond with residents of the other countries. Finally, this study provides insights into how technology can be used for learning social studies outside of the school walls.

Research question 1: In what ways do students develop global literacy as the result of blogging with Swaziland and Malaysian residents? Over the two weeks, students exhibited signs of developing knowledge about the target countries. Most participants began with simplistic stereotypes and misguided notions of controversial issues, and were able to develop sophisticated understandings through discourse with the cultural experts. The answers suggested that students gained the greatest amount of global literacy in state of the planet awareness. For example, one student stated in the final interview, "I didn't even know Malaysia existed until the blog activity," indicating both little initial knowledge as well as a developing state of the planet awareness. Another student demonstrated knowledge of global dynamics when saying, "Now I understand why some Americans travel to countries that are in need of medical attention because of the lack of education most of the workers have, and I know their [sic] trying their best to provide care for the pt" (CD, Blog post 6, May 27, 2011).

The analysis revealed an increased progression in global literacy throughout the blogging exercise (see Table 2). Further, it also provided evidence of a correlation between depth of global literacy answers and increased instances of dialogue.

Table 2. Emergent Themes from Global Literacy in Blog Prompts (M=Malaysia; S=Swaziland)

Prompt	State of the Planet Awareness	Knowledge of Dynamics	Awareness of Human Choices	Media Literacy	Research & Thinking Skills	Global Literacy
1						
2	M, S					
3	M, S	M				
4	M, S					
5	M, S	M				
6	M, S	M, S				

Table 3 provides examples of ways in which students voiced a more profound knowledge of state of the planet awareness after completing the blog activity. By the end of the activity, while many students still focused on declarative knowledge, some were able to recognize consequences of the gained knowledge, for example, the influence that religion has on culture.

Research question 2: To what extent is student cross-cultural awareness enhanced as a result of the blogging experience? The participants began the study with little or erroneous knowledge about the target cultures. The data suggest that students developed some measure of an increased cross-cultural awareness. When investigators coded the data, all participants were initially at the first level of cross-cultural awareness, "awareness of superficial or visible traits; stereotypes." While no student responses demonstrated a level 4 awareness, which Hanvey (1976) argues normally takes years to attain, by the conclusion of the exercise many student answers suggested a level 2 awareness and a few provided insights conveying a flirtation with level 3, when the person begins to intellectually process contradictions to previously held stereotypes. The following post provides an example of this increase in cross-cultural awareness levels:

Table 3. Most Common Themes from Global Literacy in Swaziland and Malaysia Questionnaires in Order of Most to Least Prominent

Sub-Dimension	Quest 1	Quest 2	Quest 3
State of the planet awareness	"I don't know"	Swaziland: HIV/AIDS prevalent. Inside South Africa	Swaziland: males dominate, low life expectancy, access to technology.
		Malaysia: multiethnic Islam, Christian, Hindu SE Asia	Malaysia: multiethnic, religion influential, mountains, tropical, modern urban cities
Knowledge of global dynamics	not addressed (n/a)	n/a	n/a
Awareness of human choices	n/a	n/a	n/a
Media literacy	n/a	n/a	n/a
Research & thinking skills	n/a	n/a	n/a

> I would just like to thank you for your responses. They were very helpful and I learned about things that I had no idea about before, like Malaysia having an active night life. When I first heard of Malaysia I perceived it as poor, uneducated people that didn't have very advanced technology systems. I have now learned that I was far from right. (IJ, Blog post 5, May 25, 2011)

Table 4 provides additional examples of student responses that show an evolution from low to higher levels of cross-cultural awareness. While individual data sources could not determine increases by themselves, when combined with remaining data sources, cross-cultural awareness increases emerged.

Recommendations

The blogging activity provided students an engaging authentic learning experience with course concepts. A drawback to this activity may be that it took place at the end of the academic year when students often begin to mentally disengage. Most students voiced enjoyment in the activity and were able to demonstrate slight increases in cross-cultural awareness and global literacy while

Table 4. Emergent Cross-cultural Awareness Categories and Demonstrated Examples in Questionnaires & Blogs

Category	Examples
1. Stereotypes	Dark-skinned and skinny (Swaziland Questionnaire 1) Living in huts (Malaysia Questionnaire 1)
2. Awareness of contrasts in cultural traits across conflicting situations	"I find it interesting that teens over there have much in common with us over here in the States in terms of activity, etc." (EF, Swaziland Blog 3) "Everything you do there is basically the same things we can do here in the United States. . . . But it is cool sometimes when you are under age and you can drive or paintball also being able to bribe officers so you won't get a ticket. That's cool." (GH, Malaysia Blog 2)
3. Intellectual analysis of of differing cultural traits	"I never considered the conservative nature the Muslim religion to have an effect on the exposure of music and fashion, but that makes perfect sense." (EF, Malaysia Blog 6)
4. Gaining insider's point of view	n/a

blogging with cultural experts over a short two-week period; therefore, doing the activity over the course of the school year would allow students to apply key AP concepts through an authentic learning experience and allow the instructor to assess students' knowledge of the concepts while continuously providing the students opportunities to apply the course material. The longer duration would also provide students the opportunity to more likely develop deeper understandings of global literacy and cross-cultural awareness. In addition, it is recommended that this study be replicated by conducting this activity over the course of the entire school year.

References

Banks, J. A., McGee Banks, C. A., Cortes, C. E., Hahn, C. L., Merryfield, M. M., Moodley, K. A., et al. (2005). *Democracy and diversity: Principles and concepts for educating citizens in a global age*. Seattle: University of Washington, Center for Multicultural Education.

Carano, K. T., & Berson, M. J. (2007). Breaking stereotypes: Constructing geographic literacy and cultural awareness through technology. *The Social Studies*, 98(2), 65-70.

Carano, K. T., Keefer, N. E., & Berson, M. J. (2008). Youth talk for civic action: An examination of digital networking tools. In P. J. VanFossen & M. J. Berson (Eds.), *The electronic republic? The impact of technology on education for citizenship* (pp. 56-74). West Lafayette, IN: Purdue University Press.

College Board. (2011). *Human geography: Course description*. Retrieved from http://apcentral.collegeboard.com/apc/public/repository/ap-human-geography-course-description.pdf

Cortes, C. E. (2000). *The children are watching: How the media teach about diversity*. New York: Teachers College Press.

Evangelista, B. (2010, February 15). Facebook directs more online users than Google. *San Francisco Chronicle*. Retrieved from http://articles.sfgate.com/2010-02-15/business/17876925_1_palo-alto-s-facebook-search-engine-gigya

Ferriter, W. M. (2010). How flat is your classroom? *Educational Leadership*, 67(7), 86-87.

Gaudelli, W. (2006). Convergence of technology and diversity: Experiences of two beginning teachers in web-based distance learning for global/multicultural education. *Teacher Education Quarterly*, 33(1), 97-116.

Hanvey, R. G. (1976). *An attainable global perspective*. New York: American Forum for Global Education.

Heilman, E. E. (2009). Terrains of global and multicultural education: What is distinctive, contested, and shared? In T. F. Kirkwood-Tucker (Ed.), *Visions in global education: The globalization of curriculum and pedagogy in teacher education and schools* (pp. 25-46). New York: Lang.

Imbert, P. (2004). Globalization and difference: Displacement, culture and homeland. *Globalizations*, 1, 194-204.

Joftus, S. (2004). *High schools for the new millennium: Imagine the possibilities*. Redmond, WA: Bill and Melinda Gates Foundation. Retrieved from http://www.edstrategies.net/files/gatesedwhitepaper.pdf

Kirkpatrick, D. (2006). Life in a connected world. *Fortune*, 154(1), 98-100.

Kirkwood, T. F. (2001). Preparing teachers to teach from a global perspective. *The Delta Gamma Kappa Bulletin*, 67(2), 5-12.

Kirkwood-Tucker, T. F. (2009a). From the trenches: The integration of a global perspective in curriculum and instruction in the Miami-Dade County public schools. In T. F. Kirkwood-Tucker (Ed.), *Visions in global education: The globalization of curriculum and pedagogy in teacher education and schools* (pp. 137-162). New York: Lang.

Kirkwood-Tucker, T. F. (2009b). Tales from the field: Possibilities and processes leading to global education reform in the Miami-Dade County public schools. In T. F. Kirkwood-Tucker (Ed.), *Visions in global education: The globalization of curriculum and pedagogy in teacher education and schools* (pp. 116-136). New York: Lang.

Landorf, H., & Nevin, A. (2007). Inclusive global education: Implications for social justice. *Journal of Educational Administration*, 45, 711-723.

Merryfield, M. M. (2000). Why aren't teachers being prepared to teach for diversity, equity, and global interconnectedness? A study of lived experiences in the making of multicultural and global educators. *Teaching and Teacher Education*, 16, 429-443.

Merryfield, M. M. (2001). Moving the center of global education: From imperial world views that divide the world to double consciousness, contrapuntal pedagogy, hybridity, and cross-cultural competence. In W. B. Stanley (Ed.), *Critical issues in social studies research for the 21st century* (pp. 179-207). Greenwich, CT: Information Age.

Merryfield, M. M., & Wilson, A. (2005). *Social studies and the world: Teaching global perspectives.* Silver Spring, MD: National Council for the Social Studies.

Metz, C. (2006). MySpace nation. *PC Magazine, 25*(12), 76-80, 83-84, 87.

Miles, M. B., & Huberman, A. M. (1994). *Qualitative data analysis* (2nd ed., pp. 10-12). Newbury Park, CA: Sage.

National Council for the Social Studies. (2010). *National curriculum standards for social studies: A framework for teaching, learning, and assessment.* Silver Spring, MD: Author.

Pike, G., & Selby, D. (2000). *In the global classroom–2.* Toronto, Ontario, Canada: Pippin.

Tye, K. A. (1999). *Global education as a worldwide movement.* Orange, CA: Interdependence Press.

Zong, G. (2002). Can computer mediated communication help to prepare global teachers? An analysis of preservice social studies teachers' experience. *Theory and Research in Social Education, 30,* 589-616.

Zong, G. (2009). Global perspectives in teacher education research and practice. In T. F. Kirkwood-Tucker (Ed.), *Visions in global education: The globalization of curriculum and pedagogy in teacher education and schools* (pp. 71-89). New York: Lang.

Appendix A

Questionnaire 1: Swaziland

Name:_____

Please answer the following questions. This questionnaire will provide us with information that will help us develop the cultural curriculum for the course. Please be honest, since this questionnaire will not affect your grade.

I. On a scale from 1 (strongly agree) to 5 (strongly disagree), answer the following statements.

1. Given that I am taking AP Human Geography, it is relevant to speak with Peace Corps volunteers in Swaziland.

   1   2   3   4   5

   Explain: _____

2. I believe that all people in Swaziland have the same values towards family, money, education, and religion.

   1   2   3   4   5

   Explain: _____

3. I believe that talking to my blog partner(s) will help me see that there are differences between people from the United States and people from Swaziland

   1   2   3   4   5

   Explain: _____

II. Definitions. Define and describe the following:

1. Culture of Swaziland:

2. Religion of Swaziland:

3. Cultural Landscape of Swaziland:

4. Sequential Occupation of Swaziland:

5. A Person from Swaziland:

6. Standard of Living in Swaziland:

7. Location of Swaziland:

Questionnaire 1: Malaysia

Name:_____

Please answer the following questions. This questionnaire will provide us with information that will help us develop the cultural curriculum for the course. Please be honest, since this questionnaire will not affect your grade.

I. On a scale from 1 (strongly agree) to 5 (strongly disagree), answer the following statements.

1. Given that I am taking AP Human Geography, it is relevant to speak with students in Malaysia

   1   2   3   4   5

   Explain: _____

2. I believe that all people in Malaysia have the same values towards family, money, education, and religion.

   1   2   3   4   5

   Explain: _____

3. I believe that talking to my blog partner(s) will help me see that there are differences between people from the United States and people from Malaysia

   1   2   3   4   5

   Explain: _____

II. Definitions. Define and describe the following:

1. Culture of Malaysia:

2. Religion of Malaysia:

3. Cultural Landscape of Malaysia:

4. Sequential Occupation of Malaysia:

5. A Person from Malaysia:

6. Standard of Living in Malaysia:

7. Location of Malaysia:

Appendix B

Questionnaire 2

I. Define and describe the following:

 1. Culture of Malaysia:

 2. Religion of Malaysia:

 3. Cultural Landscape of Malaysia:

 4. Sequential Occupation of Malaysia:

 5. A Person from Malaysia:

 6. Standard of Living in Malaysia:

 7. Location of Malaysia:

II. Define and describe the following:

 8. Culture of Swaziland:

 9. Religion of Swaziland:

 10. Cultural Landscape of Swaziland:

 11. Sequential Occupation of Swaziland:

 12. A Person from Swaziland:

 13. Standard of Living in Swaziland:

 14. Location of Swaziland:

Appendix C
Questionnaire 3: Swaziland

Please answer the following questions. This questionnaire will provide us with information reading your experience with the Swaziland culture (think of your experience in the class and your conversation with the blog partners). By answering this questionnaire you will help us develop the curriculum for future classes. Please be honest, since this questionnaire will not affect your grade.

I. On a scale from 1 (strongly agree) to 5 (strongly disagree), answer the following statements.

After I took this class, I think it is relevant to learn about Swaziland culture (Swaziland).
    1   2   3   4   5
Explain: _____

2. After studying Swaziland, I believe that all people in Swaziland have the same values towards family, money, education and religion.
    1   2   3   4   5
Explain: _____

3. After talking to my blog partner(s), I have learned that there are differences between people from the United States and people from Swaziland.
    1   2   3   4   5
Explain: _____

II. Definitions:

1. In general terms, define what you learned about the culture of Swaziland.
    a. In class
    _____
    b. From blog partner(s) comment/questions
    _____

2. In general terms, define what you learned about the religion of Swaziland.
    a. In class
    _____
    b. From blog partner(s) comment/questions
    _____

3. Define what you learned about the cultural landscape of Swaziland.
    a. In class
    _____
b. From blog partner(s) comment/questions
    _____

4. Define what you learned about the Sequential Occupation of Swaziland.
    a. In class
    _____
    b. From blog partner(s) comment/questions
    _____

5. Define what you learned about a person from Swaziland.
   a. In class
   _____
   b. From blog partner(s) comment/questions
   _____
6. Define what you learned about the standard of living in Swaziland.
   a. In class
   _____
   b. From blog partner(s) comment/questions
   _____
7. What additional aspects of the culture have you learned from your blog partner(s)?
   _____

III. Questions about blog partners:

1. Think about the topics you discussed with your blog partners. Do/does any of the comments in the blogs make you see a perspective of the United States that you had not considered before? Explain.
   _____
2. Have your previous ideas about Swaziland changed now that you have talked to your blog partners? Explain.
   _____
3. What aspects of the culture from Swaziland are you still curious about?
   _____

Questionnaire 3: Malaysia

Please, answer the following questions. This questionnaire will provide us with information reading your experience with the Malaysian culture (think of your experience in the class and your conversation with the blog partners). By answering this questionnaire you will help us develop the curriculum for future classes. Please, be honest, since this questionnaire will not affect your grade.

I. On a scale from 1 (strongly agree) to 5 (strongly disagree), answer the following statements.

After I took this class, I think it is relevant to learn about Malaysian culture (Malaysia).
   1   2   3   4   5
Explain: _____

2. After studying Malaysia, I believe that all people in Malaysia have the same values towards family, money, education and religion.
   1   2   3   4   5
Explain: _____

3. After talking to my blog partner(s), I have learned that there are differences between people from the United States and people from Malaysia.
   1   2   3   4   5
Explain: _____

II. Definitions:

4. In general terms, define what you learned about the culture of Malaysia.
   a. In class
   _____
   b. From blog partner(s) comment/questions
   _____

5. In general terms, define what you learned about the religion of Malaysia.
   a. In class
   _____
   b. From blog partner(s) comment/questions
   _____

6. Define what you learned about the cultural landscape of Malaysia.
   a. In class
   _____
   b. From blog partner(s) comment/questions
   _____

7. Define what you learned about the Sequential Occupation of Malaysia.
   a. In class
   _____
   b. From blog partner(s) comment/questions
   _____

8. Define what you learned about a person from Malaysia.
   a. In class
   _____
   b. From blog partner(s) comment/questions
   _____

9. Define what you learned about the standard of living in Malaysia.
   a. In class
   _____
   b. From blog partner(s) comment/questions
   _____

10. What additional aspects of the culture have you learned from your blog partner(s)?
    _____

II. Questions about blog partners:

11. Think about the topics you discussed with your blog partners. Do/does any of the comments in the blogs make you see a perspective of the United States that you had not considered before? Explain.

12. Have your previous ideas about Malaysia changed now that you have talked to your blog partners? Explain.
    _____

13. What aspects of the culture from Malaysia are you still curious about?
    _____

Appendix D

Interview Questions
1. Did you carefully read the blog postings? Did you find them interesting? Why?
2. After reading and posting to the blogs, how much time did you (approximately) spend thinking about what you had read afterward?
3. Did you discuss the blog postings with your classmates on your own time? If yes, what did you talk about?
4. What development was the most surprising? Why?
5. What was the most controversial blog you read? Why?
6. What did the blogging activity allow you to do that you don't think could be done another way?
7. Do you view your own culture differently because of this experience? If yes, what is an example?
8. How was this experience the same as reading articles or the textbook? How was it different? Which do you prefer? Why?
9. If you were to do this again, what would you do differently? What should the teacher do differently?
10. Were your questions answered to your satisfaction? Please explain.

CHAPTER SIXTEEN

Using Storytelling and Drama to Teach Understanding and Respect for Global Values and Beliefs

Thomas N. Turner
Dorothy Blanks
Sarah Philpott
Lance McConkey

The essence of global education is building understanding of the complex and interrelated cultural, economic, political, and environmental systems of our world. The primary goal of global education is to prepare students to be effective and responsible citizens in a global society (Masataka & Merryfield, 2004). Global education needs to be a dimension that courses through the curriculum and a way of approaching everything we teach (Global Teacher Project, 2012). It implies the study of different countries and cultures and the problems and issues that face them. Drama and folk stories provide ideal teaching mechanisms through which to approach global education. Almost every type of dramatic activity and every story of and from a particular culture are rife with opportunities for evoking empathy and sympathy, enhancing admiration and problem solving, and sparking motivational emotion grabbing.

Gay and Hanley (1999) identified four ideas in social studies that are conducive to developing multicultural empowerment for middle school students. These ideas are equally relevant for global education and for students of all ages, and include the following: the importance of civic participation, which means active involvement and the possible reform of social, political, and economic institutions; developing a sense of community membership by realizing that humans are interdependent and interconnected; cooperation and collaboration, which grow when we work together for the common good, and share power, resources, privileges, and responsibilities; and the ability to problem solve and make decisions. The development of a solid understanding of and belief in these ideas, along with the critical thinking and analytical abilities they encompass, is central to being a good world citizen and for human life in general.

In this chapter the authors discuss how global storytelling and drama relate in very intrinsic ways to global education. They also relate these teaching tools and global education to technology.

Storytelling

Stories and their telling have always been powerful, dynamic, and memorable tools for transmitting values and history within cultures and across cultures. They provide an intimate, humanizing perspective into the lives of historic and contemporary, real and imaginary people. This is why storytelling of one sort or another has been so important to the social studies. As stated in the 1994 National Council for the Social Studies Standards and echoed in the 2010 revision, "The social studies are the integrated study of the social sciences and humanities to promote civic competence" (NCSS, 1994, 2010). Civic competence today certainly refers to citizenship in a global society. The world is interconnected culturally, economically, socially, and historically. Stories can help the world become more connected in the sense of understanding different points of view. International children's stories can help transmit this knowledge because they provide young people a unique look into the lives of people across the globe.

An international children's story is a book written from an authentic perspective. This means that someone from within the culture created or captured the story and placed it into a shareable format. These types of stories can provide readers and listeners a unique perspective into the lives of individuals within societies geographically distanced from their own. Buck (2008, p. 101) has argued that "international children's texts open up the doors to new and exciting places for students to explore and for that reason they are a powerful resource for teachers. Because the texts originate in different cultures with different political structures, religions, morals, and customs, the texts will certainly expose children to lifestyles very different from their own." Opening these doors to students is one way that social studies educators can introduce students to issues of globalization at an early age. These stories give children eyes into the lives and values of people of every culture. There are universal stories that appear in many cultures: Cinderella figures, Red Riding Hood characters, and many more. There are also stories so particular to certain cultures that they inform about the climate, the geography, the culture and cultural values, and even what people eat and drink in their cultural settings. Finally, there are stories that give breath and life to the peoples of times past.

There are uncountable ways to utilize stories. Teachers need to read aloud to students of all ages, tell stories to them in many ways, and teach them to listen to stories for the pleasure of hearing the adventures and passions of people everywhere. Stories can be read and analyzed to bring understanding of other people, other cultures, and other times. The following are just a few of the many possibilities for elementary, middle level, and secondary and college classrooms:

- Use international children's literature to show readers that life around the globe is current and ongoing. Many students seem to think that people around the globe live in antiquated societies—that they have not moved beyond what is presented in history books. International children's books can show young people the modernity of other cultures. In Péter Nyulász's (2008) *Come with Us to Budapest!*, readers travel to the city of Budapest. The young protagonist and her family explore the sites of the historic city while eating ice cream, wearing modern clothing, and riding the metro. Through looking at illustrations and reading the text we find that places we learn about in history, such as Budapest, are dynamic places that change with the times.
- Read or tell stories such as Demi's (2007) *The Empty Pot* or any of her stories about historic world leaders or world religion founders, which humanize different people of the world and make them sympathetic and likeable.
- Discuss world stories, which illustrate that although we are different in many respects, citizens in a global society possess many of the same core values. For example, *Buhuki*, written and illustrated by Janaki Sooriyarachchi (2011), originally published in Sri Lanka, is a humorous tale about a monkey who gets into trouble for not sharing his snack with others. This universal message shows readers that sharing is a valued action. Many other books illustrate shared values, fears, joys, and dreams.
- Learn some of the international stories and develop your storytelling skills to introduce students to that art. Then invite students to learn some of the stories and share them with their classmates.
- Use pointed questions that require students to analyze the illustrations in picture books about different cultures to look at the geography, architecture, food, tools, and clothing of other places. The ever-growing abundance of wonderful, authentically illustrated trade books for all levels makes this an easy task.
- Use books to initiate conversations that are difficult to tackle because of preformed student opinions or controversy. Conversations about books provide a safe place for students to tackle complex issues (Brooks & Hampton, 2005).
- Invite storytellers from other cultures to share the storytelling methods and traditions of their homelands.

Integrating international children's literature into the classroom should be done with purpose and discretion. Books should be selected on the basis of cultural authenticity and be free of stereotyping. Some international children's books might appear controversial to young readers. Buck (2008) found that students frequently make value judgments when reading international children's books. She went on to say that "teachers and librarians must carefully review the literature and "consider how to handle discussions of sensitive topics concerning lifestyles, religions, and cultures before they are encountered within the classroom" (p. 101). By anticipating these conversations educators can lead discussions in which students can analyze and come to understand that people possess different perspectives all across the globe.

Finding stories that retain the cultural integrity of the text when translated may be difficult. The United States Board on Books for Young People (USBBY) releases an annual list of outstanding international books. Each entry includes the title, country of origin, and recommended grade level. The International Board on Books for Young People (IBBY) also publishes a list of recommended books from around the world. The International Children's Library is another valuable resource. It has transferred many award-winning books into a free digital format. NCSS also provides another list of notable trade books every year, many of which deal with global issues and people from a variety of world cultures (NCSS, 2012).

Considering Readability

The readability of the books and stories you use is an essential consideration, and that involves both reading level and the cultural load of the text. Buck (2008, p. 3) points out that "teachers also need to consider that students might not come equipped with the appropriate background and cultural understanding." This really should not limit the selection of books so much as it should challenge the teacher to find creative ways of building background.

Drama to Promote Global Education

Like storytelling, drama is a multidimensional, holistic approach to learning that evokes and extends students' intellectual, social, emotional, physical, moral, creative, communicative, and aesthetic abilities (Verriour, 1994). It involves a spectrum of types of activities, including short plays, readers' theater, simulations, mock trials, reenactments, role-playing, choral reading, and other strategies that draw students into the action of their own learning (Turner,

2004). The integration of drama into social studies has benefits for both social science and language arts. Reading and writing skills are developed through historical dramas (Howlett, 2007). Students who act out history enhance their understanding of historical content (Howlett, 2007; Kelin, 2005; Mattioli & Drake, 1999).

Various types of drama have been effective in social studies classrooms. Process drama is one technique that was found to encourage students to empathize with students who were different from themselves (Rosler, 2008). Obenchain and Morris (2001) used melodrama, and they reported that students learned about characters from a particular historical era and not only examined but also actually took on their values. Morris (2002) coupled drama and primary sources into a class soap opera, "As the Civil War Turns." He reported that this method provided students with engaging, relevant, and active learning experiences. Kelin (2005) described third-grade students acting out Henry Hudson's voyages of discovery. He reported that the lessons elicited genuine emotional responses in students.

Drama can have major benefits for global education. Student perceptions are changed and a sense of connection with others is developed, and they approach situations from multiple points of view. They think more critically, and examine issues, ideas, values, and perspectives from the safety of the drama. Morris (2003) concluded that through plays students studied questions raised in the past and applied them to current social problems. Students thought critically about interconnections between well-known people and important issues. Obenchain and Morris (2001) also spoke about connectedness; they found that the use of melodramas had students addressing multiple perspectives, examining values, and making historical connections across time and space.

Chilcoat (1995) found that through drama students explored the meaning of self-determination and activism, and that the lessons were a forum for discussion of real problems and a means for social action. Kelin (2005) had upper elementary students use dramatic role-playing and improvisation to explore the American decision to use the Bikini Atoll for nuclear testing. Through drama, students can practice skills of community, participation, cooperation, collaboration, problem solving, and decision making (Gay & Hanley, 1999). Drama activities involve working together as a community, creating a single product—a performance. Drama itself is inherently hands-on and "minds-on," whether involving dramatic reading, staging a play, or doing any of the spectrum of activities in between. Specific drama activities, such as structured role-playing, psychodrama, cliffhanger readings (Turner, 2004), and Storypath (McGuire & Cole, 2005) are themselves problem centered with post-drama discussion at their heart. Most drama requires decision making and certainly collaboration. Drama is, in many ways, the perfect teaching approach to address all four of these important concepts.

Dramatic activities can be used to provide cross-cultural experiences, which are described by Masataka and Merryfield (2004) as a major strategy to help young people to understand their globally interconnected world. Role-playing and simulation offer unlimited possibilities. Students can role-play problem solving and cultural contacts among people of different cultures, different times, and different places. Simulations are the most well-known strategies. Students worldwide have been involved in Model United Nations simulations. Student ambassadors gain new knowledge and skills through debate and deliberation, looking for change by providing solutions with a focus on human rights, global stability, and peace-building competencies (Kirkwood-Tucker, 2004). Another simulation centered on world economics is the Trading Game (2012), in which students try to make as much money as possible in a limited amount of time. They experience problems of free trade and increased understanding of economic globalization (Wilkinson, 1985). The best approaches, of course, are teacher-developed simulations.

Readers' theater has gained many advocates over recent years. Readers' theater allows students to experience multiple perspectives. Maher's (2006) *Most Dangerous Women: Bringing History to Life through Readers' Theater* meticulously outlines how teachers can incorporate readers' theater in their classrooms in a way that makes history come alive while teaching for positive peace. In Curran's (1996) *Peace Pilgrim: A Readers' Theatre Approach to Peace Education*, pre-service teachers first read a biography of the "Peace Pilgrim" and then wrote and performed a readers' theater play centered on the ideas of peace education. Salter (1992) used a readers' theater script titled "The Other Side of Discovery," about the effects of Columbus's arrival. The script asked students to think more deeply about the traditional Columbus story. "When Iron Crumbles: Berlin and the Wall" was used in a secondary classroom to explore the history of the wall in Berlin (Chan, 1991). The readers' theater script "Collapse of a Multinational State: The Case of Yugoslavia" was used with college students to introduce events leading to World War I (Steinbeck, 1994). These examples demonstrate the breadth of topics and the age range that can be involved in readers' theater.

The imaginative teacher can create a variety of relevant, engaging, drama-based lessons related to global education. The wide range of dramatic activities includes melodrama and soap operas, structured role-playing, socio-dramas, docudramas, mock trials, simulations, and readers' theater. Pantomime, mock television commercials, mockumentaries, and music videos also are among the possibilities. Environmental global issues could be explored, for example, in a mock trial of the *Exxon Valdez* crew, the destroyers of the rain forests, or poachers of endangered elephants and rhinos. Classroom debate can be held between countries such as Pakistan and India on the need for nuclear weapons. Reenactments can be planned and staged of civil rights marches or Gandhi-like nonviolent protests. There are almost unlimited combinations of drama strategy, historical, and global education concepts that can be imagined in powerful lessons for students of all ages.

Technology as a Tool for Storytelling and Drama in Global Education

Technology is at once a tool of access to drama and storytelling examples, material, and information, and a stage where students can enact drama and storytelling. Imagine a classroom where students are reading stories from ancient Egypt, writing scripts, and preparing to use their cell phones or iPods to create a digital version of the story they just read. No longer a dream, this is today a real possibility because almost 98 percent of U.S. public schools have high-speed Internet connections (National Center for Educational Statistics, 2010). They can call across national and cultural borders to enhance cultural and language learning.

The Internet has evolved to include digitized versions of primary sources, stories, and other documents. Teachers can find anything from a copy of the Declaration of Independence to Plato's *Republic*. Finding global sources to use in drama and storytelling projects is just a mouse click away. Powerful search engines can help teachers scan millions of sites to discover a story from another culture or a new Web tool.

Use the stories and information from Internet searches as the base from which students can create a video retelling of the story. For example, when teaching a unit on ancient Egypt, students are placed in groups and given several stories from Egyptian history including "The Girl in the Rose Red Slipper," "The Crocodile Story," and "The Doomed Prince." After they have read each story, the group selects one to use for their digital recreation. They then create a script and a storyboard to guide them in their digital retelling of the story. Students then can use cameras from cell phones, iPods, or iPads, or a video camera to record the story and transfer it to a computer for editing. Windows Moviemaker, Photostory, and iMovie are all types of video editing software.

Technology provides not only countless tools and resources for storytelling and drama, but it also provides almost limitless resources for researching and evaluating sources. We can find speeches to dramatize, photographs and paintings that can be pantomimed and discussed, contradicting opinions and alleged facts—practically anything we want our students to read and discuss.

Conclusion

If global education is to succeed it must entrance and enrapture students. Students must become and remain active in their learning about issues, people, and places worldwide. They must embrace other people and their needs with empathy and understanding. Drama and storytelling combined with the right technologies can provide exciting avenues for this very miraculous transformation to occur.

References

Brooks, W., & Hampton, G. (2005). Safe discussions rather than first-hand encounters: Adolescents examine racism through one historical fiction text. *Children's Literature in Education*, 36(1), 83-98.

Buck, C. (2008). *Young readers respond to children's literature.* Unpublished doctoral dissertation, University of Tennessee, Knoxville. Retrieved from http:// trace.tennessee.edu/utk_graddiss/332/

Chan, A. (1991). *When iron crumbles: Berlin and the wall. A social studies unit recommended for grades 9-12 and community college.* Stanford, CA: Stanford Program on International and Cross Cultural Education.

Chilcoat, G. W. (1995). Using panorama theater to teach middle school social studies. *Middle School Journal*, 26, 52-56.

Christensen, L. M. (2001). Searching for the heart of the USA: A geographical drama. *Social Studies and the Young Learner*, 14(1), 1-4.

Curran, J. M. (1996, April). *Peace pilgrim: A readers' theatre approach to peace education.* Paper presented at the annual meeting of the American Education Research Association, New York, NY.

Demi. (2007). *The Empty Pot.* New York: Holt.

Gay, G., & Hanley, M. S. (1999). Multicultural empowerment in middle school social studies through drama pedagogy. *The Clearing House*, 72(6), 364-370.

Global Teacher Project. (2012). Retrieved from http://www.globalteacher.org.uk/global_ed.htm

Howlett, C. F. (2007). Guardians of the past: Using drama to assess learning in American history. *Social Education*, 71(6), 304-307, 331.

Kelin, D. A. (2005). Voyages of discovery: Experiencing the emotion of history. *Social Studies and the Young Learner*, 23, M4-M8.

Kirkwood-Tucker, T.F (2004). Towards a theory of world-centered citizenship education. *The Development Education Journal*, 10 (2), 14-15.

Maher, J. (2006). *Most dangerous women: Bringing history to life through readers' theater*. Portsmouth, NH: Heinemann.

Masataka, K., & Merryfield, M. M. (2004). How are teachers responding to globalization? *Social Education, 68*(5), 354.

Mattioli, D. J., & Drake, F. D. (1999). Acting out history: From the ice age to the modern age. *Social Studies and the Young Learner, 11*(3), M9–M11.

McGuire, M. E., & Cole, B. (2005) Using Storypath to give young learners a start. *Social Studies and the Young Learner, 18*(3), 20–23.

Morris, R. V. (2002). Use primary sources to develop a soap opera: As the Civil War turns. *The Social Studies, 93*(2), 53–56.

Morris, R. V. (2003). Acting out history: Students reach across time and space. *International Journal of Social Education, 18*(1), 44–51.

National Center for Education Statistics. (2010). *Digest of education statistics*. Retrieved from http://nces.ed.gov/programs/digest/d10/tables/dt10_108.asp

National Council for the Social Studies. (1994). *Expectations of excellence: Curriculum standards for social studies*. Washington, DC: Author.

National Council for the Social Studies. (2010). *National curriculum standards for social studies: A framework for teaching, learning, and assessment*. Washington, DC: Author.

National Council for the Social Studies. (2012). *Notable trade books for young people*. Retrieved from http://www.socialstudies.org/resources/notable

Nyulász, P. (2008). *Come with us to Budapest!* (P. Mullowney, Trans.). Hungary: Color Plus Kft. Retrieved from http://www.childrenslibrary.org/icdl/BookPreview?bookid=nyugyer_00740022&route=text&lang=English&msg=&ilang=English

Obenchain, K. M., & Morris, R. V. (2001). The (melo)drama of social studies: Our hero! *The Social Studies, 92*(2), 84–85.

Rosler, B. (2008). Process drama in one fifth-grade social studies class. *The Social Studies, 99*(6), 265–272.

Salter, C. L. (1992). The other side of discovery. *Journal of Geography, 91*(1), 11–17.

Schuchat, D. (2005). Radio days in the classroom. *Social Education, 23*, M4–M8.

Sooriyarachchi, Janaki. (2011). *Buhuki*. Bibliotastic. Retrieved from http://www.bibliotastic.com/ebooks/fiction-juvenile/buhuki

Steinbeck, R. (1994). *Collapse of a multinational state: The case of Yugoslavia. A curriculum unit for history and social studies*. Stanford, CA: Stanford Program on International and Cross Cultural Education.

The trading game. (2012). econedlink. Retrieved from http://www.econedlink.org/lessons/index.php?lid=855&type=educator/

Turner, T. N. (2004). *Essentials of elementary social studies*. Boston: Pearson.

Verriour, P. (1994). *In role: Teaching and learning dramatically*. Toronto, Ontario, Canada: Pippin.

Wilkinson, A. (1985). *It's not fair*. London: Christian Aid.

CHAPTER SEVENTEEN

World Tour by Bus: Teaching and Learning about Globalization by Exploring Local Places in Search of Global Connections

Aaron T. Bodle

Global educators are faced with a conundrum. How can we support students' acquisition of skills, knowledge and attitudes ideal to global citizenship when, for most students, "global" is conceptualized as something distant, abstract and removed from their lived experiences? Furthermore, curricular materials used in global education often theorize the global as distant and portray a false dichotomy between local communities and global processes (Merryfield, 1996). Yet one key goal of global education is to promote awareness of the implications of our local actions for the greater global world (Anderson et al., 1994; Hanvey, 1976). If global education is to fulfill this goal, teachers and curriculum designers must connect the local with the global in an effort to make both more meaningful as they are enacted across the curriculum (Merryfield, 1996; National Council for the Social Studies, 2001; Noddings, 2005). I argue this is particularly necessary in social studies classrooms.

This chapter, based on data from an ethnographic study conducted in 2011, highlights the efforts of a pair of dedicated global educators who have attempted to address these concerns about global education by taking their American history students directly to the places and spaces where the global meets the local in their Rust Belt community of what I call Factory Town. Believing, as Noddings does, that one way to connect global and local interests is to ask students to "study local places appreciatively and communicate something about them to the larger world" (Noddings 2005, p. 62), these teachers organize a yearly two-hour field trip around the local community. The field trip is intended to support students' learning about connections between their community and global processes, places and people.

"Placing" Factory Town

All spaces, according to Doreen Massey, are both local and global. Like the relational construction of an individual's identity, places and spaces are produced through social relations. Webs of discursive power relations are produced concomitantly and relatedly within global and local spaces. These spaces are produced through

> practices, trajectories, interrelations. . . . If we make space through interactions at all levels, from the (so-called) local to the (so-called) global, then those spatial identities such as places, regions, nations, and the local and the global, must be forged in this relational way too, as internally complex, (and) essentially unboundable in any absolute sense. (Massey, 2004, p. 9)

In other words, what we perceive as separate local and the global spaces are, in fact, always in relation with no real boundaries. It is important to note that although I run the risk of perpetuating the false dichotomy between the "local" and the "global," I use the terms separately in this chapter. I chose to separate the two terms here in order to discuss the processes of globalization in an intelligible way, exploring how seemingly global processes such as globalization have an impact upon seemingly local spaces and vice versa. I now turn to a description of Factory Town, which will place the community within simultaneously global and local spheres of relations.

Global processes have had a profound influence on Factory Town since its earliest settlement. The Native American tribes who inhabited the area now known as Factory Town were part of global trade routes that reached as far as Central America via the Great Lakes, Mississippi River and the Gulf of Mexico. Arriving in the seventeenth century, French explorers settled in the area and developed a fur-trading empire, supplying furs to hat makers throughout North America, Europe and Asia. By the late nineteenth century Factory Town was quickly becoming an important industrial city with a developed lumber trade and an expanding manufacturing sector based on the production of carriages. In the early twentieth century the carriage industry gave way to the mass production and distribution of automobiles and auto parts. Throughout each economic era of the city, regional, national and global demands drove production of these goods. As productivity increased, the size of the city and its economic fortunes grew—that is, until the oil embargoes of the 1970s drove demand for fuel-efficient cars built in Japan and Korea, and the subsequent expansion of neoliberal economic policies in the 1980s and 1990s. These global events led to a massive contraction of the U.S. automobile industry, leading to a rapid process of deindustrialization that has characterized Factory Town ever since.

According to the U.S. Bureau of the Census records, in 1970 Factory Town was home to roughly 200,000 residents. Most of these residents held steady blue-collar jobs in the robust automobile sector (Department of Technology, Management and Budget, 2010). Since deindustrialization, some estimate that Factory Town has lost over 60,000 auto industry jobs, and untold numbers of manufacturing jobs with parts suppliers and in other sectors of the secondary economy. As a result, Factory Town lost over 100,000 residents, down 50 percent over the past four decades (Department of Technology, Management and Budget, 2010). Currently, 32 percent of all land parcels in Factory Town are abandoned. In 2008, only six single-family homes received permits for construction while countless abandoned homes were demolished. Huge parcels of land that once housed the world's largest manufacturing facilities stand empty and overgrown with scrub vegetation surrounding their rusted-out infrastructure, light-posts, curb-bumpers and asphalt.

While economic globalization and its subsequent dependent deindustrialization have recently had devastating effects on Factory Town, the community remains culturally and socially rich. Although its crime rate remains high and the prospect for employment for a large portion of the population of the city remains low, many Factory Towners maintain a strong sense of pride in the community's history and many retain high hopes for the future. A downtown urban renewal effort is underway, multiple highly respected art and social museums continue to thrive and a highly active civil society is maintained through neighborhood groups and other nongovernmental organizations. Factory Town is also at the forefront of the growing urban agriculture movement, which is repurposing empty urban spaces for sustainable agricultural use. Despite its economic challenges, many in Factory Town see it as a thriving community, a place that is on the cutting edge of sustainable urban renewal. The two social studies teachers whose efforts are highlighted below are two of Factory Town's most hopeful and dedicated residents.

The Teachers

Ellen Miller and Danny Carpenter are co-teachers and designers of an American history course centered on social history, critical democratic citizenship and global education. Inspired teachers, Danny and Ellen work closely with their students in an alternative high school that serves a racially and economically diverse group of students who have had limited success in the traditional public school setting. Going against prescribed state standards, these teachers combine American history with world history in an attempt to complicate what they see as a flawed division between the subjects. Danny explains his and Ellen's rationale for combining subjects as follows: "What we do is, we say there is history and we are going to deal with it as history [Danny's emphasis]. So when we deal with American history in terms of Factory Town, we deal with it with the international aspects immediately." This is an ambitious task, however, and they recognize that the content they hope to engage using this approach must be grounded for students in ways that are meaningful in their everyday lives. Thus, Ellen and Danny have designed a brief field trip and the activities that surround it with the intent of helping students connect their community, in its present state, to processes related to global interconnectivity and globalization as they have occurred across multiple historical eras. They believe this approach will make the content meaningful by drawing geographic and temporal connections between their local context and global processes.

Both teachers are natives to Factory Town. They, like the parents and grandparents of their high school students, have a living memory of the dramatic changes in the local economy that resulted from the globalization of labor, oil embargoes and the global expansion of neoliberal economic policies in the past three decades. They consistently draw upon their "life's curriculum" (Clandinin & Connelly, 2000) to develop lessons that are steeped in social studies content regarding economic, cultural and political globalization but are attentive to the particularity of how these global concepts manifest within local spaces and places (Hall, 1991). The tour and the assignment to which it connects were born out of these teachers' experiences in the community and the recognition that their students have difficulty connecting their community to the global processes that have influenced it throughout history.

The Assignment and the Bus Trip

A two-hour bus trip around the community, typically scheduled in mid-September, serves as the foundational element of this globally attentive curricular approach. Its early position in the semester makes it possible for the teachers to use the trip as a reference point as they dig deeper into various historical periods throughout the course of the semester. The day before the trip, students are divided into groups and given the task of collectively developing a paper and class presentation based on their own research and their experiences on the field trip. The goal is for groups to choose one event, space or historical figure highlighted on the trip and to tie it to the regional, national and global contexts that created it. For example, if students choose to study a sit-down strike that occurred in the city's auto plants in the 1930s, then they must research the history of the strike and what led up to it as well as the national and global contexts that may have contributed to the local conditions that inspired the strike. In other words, students are challenged to place the strike within a constellation of other contextual factors that may have precipitated it at the local level.

World Tour by Bus

As students arrive on the day of the field trip, Danny hands them a thick itinerary, approximately ten pages in length, which includes the name of each stop along the tour as well as a detailed description of the site, including bibliographic references. Once students are settled aboard the bus and attendance is taken, the tour begins. Before the bus leaves the parking lot, students are directed to their itineraries; the parking lot itself has minor historical significance because it was part of the estate of a once-prominent automotive executive. Throughout the trip, the itinerary serves as a supplemental text to Danny's detailed comments and questions about the various stops. Little time is used for discussion during the trip, but a debriefing session the following day allows students the opportunity to discuss what they learned about their community and its global connections during the tour.

The itinerary is an eclectic mix of historically significant buildings and homes, former factory sites, brownfields, memorials to prominent residents in the city's past, city streets and schools. Most of the sites are associated closely with the past and present of the United Auto Workers (UAW) union and General Motors Corporation (GM). This emphasis is due largely to the central role each played in the construction of Factory Town's historical and global presence. Because the tour takes place in geographic order, not chronological order, students are expected to make their own historical connections between the stops, a process regularly scaffolded by the teachers during the debriefing session and follow-up assignments. Due to space limitations, I will highlight only two of the stops on the tour.

The Sit-Down Strike Memorial

Traveling south on the city's main thoroughfare, the bus passes by the first Chevrolet dealership to be opened outside of the factory on the city's west side. As the bus passes, Danny explains the history of the dealership and highlights the importance of the automotive industry in making Factory Town what it was at its height. Another brief stop quickly follows, this time at a parking lot that once housed a factory for the Fisher Body company, a key supplier of General Motors autobodies since its inception in 1912. This factory was a key battleground in the 1936–1937 Sit-Down Strike, an act that ultimately led to the formation of the United Auto Workers union. A short trip across the south side of town brings the bus to a stop at the Sit-Down Strike Memorial, a small sculpture park depicting the events of the strike and the contributions of the men and women who took great risks for fair wages, collective bargaining rights and safe working conditions. At this stop the students exit the bus and explore the park on their own, taking notes and writing down important dates and names to be used in their presentations later in the week. Visiting the memorial provides an opportunity for Danny and Ellen to explain shifts in global labor relations that resulted from the sit-down strikes and the formation of the UAW, which propelled the international expansion of organized labor and the subsequent formation of an international working middle class.

For many students, this trip to the memorial is their first visit. They have lived in Factory Town, some within a mile or two from the memorial, but they had never visited the site. The richness of Ellen's historical examples and Danny's global and international connections to the local context serve as powerful tools in supporting their efforts to draw students' attention to Factory Town's globally significant past and present. These examples enabled students to consider that global processes do influence their local community, and their community also influences global economic and cultural processes.

"Buick City"

After the visit to the Sit-Down Strike Memorial and a trip through the city's historical district, the bus arrives at the brownfield left by another massive GM facility. Until 1994, this site was home to the largest manufacturing facility in the world. In its prime in the early 1980s, this facility was populated, daily, by 28,000 workers. Today, little remains of the site but empty factory floors and a few buildings that are being slowly demolished. As they enter the site, the driver slows the bus to a crawl, giving Danny time to talk about the surroundings.

>Danny: What are these empty lots? Does anybody know?
>
>[Very few students raise their hands or comment.]
>
>Danny: They were factory floors. This area right here was actually the Buick world headquarters. Now it is the most polluted site in the United States. At its peak in the 1970s and 80s, this area housed 104 individual factories. So, how many do you think are still in operation?
>
>Student: Four?
>
>Danny: No, that was the case until 2004. ZERO! That's right, they are all gone. Does anybody know where the factories are now?
>
>Multiple students: China! Mexico!
>
>Danny: Well, you are right to some extent. The corporation moved the jobs to non-union states and overseas to increase their profits, and as a result these factories were no longer viable for General Motors to maintain. They began demolition in 2004 and

they are continuing to work on that project today in 2011. It is important to point out that we sometimes stop at the actions of GM, without thinking about the bigger economic picture of their decisions. That's something we will come back to time and time again this year.

As the bus continues to roll through this abandoned expanse of land, Danny stops talking. The students are quiet, looking out the windows at the wide expanse of deindustrialization. While it remains a space they have always known, it is now also a place of global significance, less an empty lot and more a historical marker. As the trip winds down and the students head back to school, Danny points out other former factory lots, museums, a local farmers' market and several small historical markers with relevance for Factory Town's history. As he does so, he connects Factory Town with the rest of the world via economics as well as through demographic shifts in the city's population and ideological shifts in governance and public policy.

In the days following the bus trip, students are given time and support to explore one aspect of the tour in great detail. They are to place the person, event or action in the context of regional, national or local contexts using sources found in the classroom, online or in local libraries. Some of the guiding questions that teachers pose for student research include the following: Why would Factory Town be an ideal place to support [fill in the blank] industry? Why was this a lucrative industry at the time? What was happening around the world that influenced the success and failure of your [person, industry, event]? How did your topic have an impact on the rest of the world? Students' answers to questions such as these are used to construct presentations about their topic and its link to global processes in the past and the present.

Global Connections

Can a single two-hour field trip in a familiar context solve the challenges of making concepts such as economic globalization meaningful to secondary students? Probably not, but it can serve as a good start. The field trip produced a space in which students could rethink their community's connections to the larger world. However, this rethinking was not sparked immediately by the trip alone. In fact, many student responses to the trip were stated in terms of the local and the past, contrary to the stated aims of the trip, which were to connect the past and the distant global to the present and local. However, the latter aims were facilitated over the course of the semester through careful scaffolding and teacher support as well as reflections on the brief yet powerful field trip. Danny explains it like this:

When we deal with American history in terms of Factory Town, we deal with it with the international aspects immediately. So it's part of the curriculum that we teach, and we expect the students to think about it in that way and it takes some time, it does. Because they are not ready to think that way [early in the class] but instead of dumbing it down or recasting it into an easier light, we expect them to work up to it, but by giving them more time and more experience with it including actually going to these places and seeing these (global/local and past/present) connections firsthand.

The teachers consistently refer to the trip throughout the course of the semester, prompting students to connect civil and labor rights movements and economic globalization to their local community. The teachers also ask them to draw local connections to global markets, systems and discourses. Janie, a working-class student and Factory Town native, referred to the immediacy of this approach to explain why she felt it was successful:

You learn so much more about everything around you than you do just reading a history book. Like . . . going around and actually seeing things that you can connect to all the stuff that you've been watching and learning about. It's just mind blowing because, in a regular class, you just read the book, do the worksheets and you're done. And it's like, okay, I read that thing on that place but I don't know where it is or anything about it and [in this class], everything we do, [the teachers] can tell you a place you can go and see it. And I love that.

Later, as she described her excitement upon learning that the Sit-Down Strike was an event with global significance, Janie explained, "I like that I can put my, my personal being in Factory Town-ness on the strike. I like to be able to know the history of where I'm at and to be able to say, you know the sit-down strike? That's like right around the corner from me. . . . Before I had [this class], I never knew."

Eventually, throughout the duration of the course, students develop deeper connections between their local communities and the larger global processes linking their local economy and neoliberal economic policies, and describing how knowledge can be distributed around the world. Several students referred to international labor movements, strikes and the globalization of manufacturing and supply chains as examples of global phenomena that connected Factory Town to a globalized world. As these connections grew, students' awareness of the economic injustice in their home community was heightened, and they could then develop a greater empathy for those facing similar injustices around the world.

Conclusions and Suggestions for Practice

When I set about writing this chapter I intended to shed light on a curricular approach that was highly effective in connecting global content in the social studies to students' lived experiences of their local communities. In addition, I hoped to illustrate the value of emphasizing the importance of "place" for teaching and learning global content knowledge by deliberately setting the chapter within a community, like other communities, that has distinct global connections, and by giving voice to teachers and students who live and work there. As Hall (1991, p. 62) observes,

> What we usually call the global, far from being something which, in a systematic fashion, rolls over everything, creating similarity, in fact works through particularity, negotiates particular spaces, and works through mobilizing particular identities and so on, so there is always a dialectic between the local and the global.

Therefore, an attentiveness to local places is paramount to how we help students develop an awareness of the global systems and contexts they share with others around the world, the impact of their local actions on the world and the influence of global phenomena on their local community. Meyers (2006) calls upon social studies curriculum makers to "consider ways that curriculum topics can support the local-global relationship as well as integrate current scholarship on globalization" (p. 389). Meyers' emphasis on local-global relationships is paramount to successful education about the topic of globalization for the simple reason that experience is local, and our students' understandings of concepts such as globalization are informed and maintained via existing power relations. Thus, a social studies curriculum must provide space for deliberation, exploration and reconsideration of globalization as it affects students' local contexts. By challenging students to engage with local spaces, and their interconnectivity with global influences, Ellen and Danny invite their students to take their first steps in understanding their own connections to global processes.

Factory Town has faced many challenges associated with globalization, and students recognize this. However, these challenges are turned into opportunities for global learning in the context of this unique process of teaching about globalization in the social studies classroom.

References

Anderson, C. C., Nicklas, S. K., & Crawford, A. R. (1994). *Global understandings: A framework for teaching and learning.* Alexandria, VA: Association for Supervision and Curriculum Development.

Clandinin, J., & Connelly, F. M. (2000). *Narrative inquiry.* San Francisco: Jossey-Bass.

Department of Technology, Management and Budget. (2010). Population of cities, towns and villages. Retrieved from http://www.gov/cgi/0,4548,7-158-54534_51713_51716-252541-,00.html

Hall, S. (1991). Old and new identities; Old and new ethnicities. In A. Smith (Ed.), *Culture, globalization and the world-system* (pp. 41–68). London: Macmillan.

Hanvey, R. G. (1976). *An attainable global perspective.* New York: Center for War/Peace Studies.

Massey, D. (2004). Geographies of responsibility. *Geografiska Annaler,* 86(1), 5–18.

Merryfield, M. M. (1996). *Making connections between multicultural and global education: Teaching educators and teacher education programs.* Available from American Association of College of Teacher Education Web site: http://aacte.org/Research-Policy/Recent-Reports-on-Educator-Preparation/

Myers, J. (2006). Rethinking the social studies curriculum in the context of globalization: Education of global citizenship in the U.S. *Theory & Research in Social Education,* 34(3), 370–394.

National Council for the Social Studies. (2001). *Preparing citizens for a global community: Position statement of National Council for the Social Studies* (Rev.). Retrieved from http://www.socialstudies.org/positions/global.

Noddings, N. (2005). Place-based education to preserve the Earth and its people. In N. Noddings (Ed.), *Educating citizens for global awareness* (pp. 57–68). New York: Teachers College Press.

CHAPTER EIGHTEEN

Teaching Social Studies from a Human Rights Perspective: Professional Development in a Context of Globalization

Rachayita Shah
Rosanna Gatens
Dilys Schoorman
Julie Wachtel

In today's era of standardization and accountability, social studies instruction is increasingly marginalized in public schools as teachers yield to pressures to increase curricular space for reading and math (Apple, 2000; Au, 2009; Burroughs, Groce, & Webeck, 2005; Stromquist, 2002). Even where it is still taught, the subject area is deemed boring (Loewen, 2008), because it is frequently focused on isolated and fragmented knowledge to be memorized, rather than on the study of meaningful, historical, socio-political phenomena that connect students to the society in which they live (Burroughs et al., 2005). As Gatens and Johnson (2011) note, current social issues should drive social studies curricula and should provide students with an understanding of democracy and an ethos of human rights. This chapter argues for framing social studies education as a process of conscientization (Freire, 1970/2000) about human rights, in which students and teachers become critically aware of social and political structures that variously provide access to who and what is included in curricula as well as the implications of this knowledge for civic engagement. Only when teachers are educated and supported in their efforts to integrate elements of critical pedagogy in social studies education can they transform themselves and their students in ways that contribute to justice and equality for all.

This chapter describes such an effort by examining a teacher professional development (PD) program for Florida teachers implementing state-mandated Holocaust education. The Holocaust Educators Summer Institute (HESI) is one of eight PD programs, partially funded by Florida's Department of Education in fulfillment of Florida's 1994 instructional mandate (http://flholocausteducationtaskforce.org/About/FloridaStatute) to teach about the Holocaust in Florida public schools. The language of the mandate encourages the application of critical pedagogy theory and practice to fulfill its goals. It explicitly links teaching and learning about the history of the Holocaust to analyses of human behavior such as prejudice, racism, and stereotyping that played an essential role in preparing German citizens to accept government-sponsored genocide against its Jewish minority. The explicitly stated purpose of the mandate is to strengthen democratic institutions by preparing students to become citizens who are able to live together in a pluralistic society.

Pedagogical Value of Holocaust Education

Shiman and Fernekes (1999) and Totten (1997) acknowledge the power of Holocaust education to move people in the directions of moral and ethical responsibilities such as protecting human dignity regardless of time or place. This is possible when the curriculum and instructional approaches allow for the analyses of historical events and policies in order to comprehend current social contexts (Lindquist, 2010), facilitate discussions in the context of critical democratic citizenship (Abovitz & Harnish, 2006; Feinstein, 2004), and encourage individuals to advocate for human rights (Berger, 2003; Bromley & Russell, 2010; Ekman, 2010).

Holocaust education, which in this case is conceptualized as a multidisciplinary undertaking, is particularly important in the context of increasing globalization. First, Germany's goal of increasing its global competitiveness at the expense of its Jewish minority presents a clear example of the ways in which the demands of globalization erode civic and human rights among less powerful groups. Second, Holocaust education, through its emphasis on cross-cultural understanding of human diversity and civic education and engagement, challenges the marginalization of social studies education in favor of achievement measured solely by scores on standardized tests, a troubling consequence of the pressures of globalization.

Professional Development as Conscientization

The HESI was designed to support in-service teachers' implementation of state-mandated Holocaust education. Dealing as it does with injustice and oppression of minorities, professional development surrounding the Holocaust necessarily engages teachers in critical reflection about themselves and their students. This critical reflection frequently transforms teachers and students from passive analysts of human behavior to engaged citizens seeking to apply knowledge to improve the human condition in the here and now. Freire (1970/2000) characterizes this transformation as a process of conscientization, in which teacher and students, through dialogic problem-posing, gain critical awareness of and become actors engaged in ameliorating social injustices. Unlike the more traditional "banking" approaches to education, characterized by memorization, student passivity, and instructor dominance—in Freire's words, a "pedagogy of oppression"—critical pedagogy both emancipates and humanizes teachers and learners.

The design of the HESI curriculum reflects principles of critical pedagogy. These include the integration of historical information with discussions of its contemporary significance; examination of topics from diverse and/or underrepresented and international perspectives (Zinn, 2007); the organization of each day's content around "big ideas" (Sleeter, 2005) central to social justice; and a focus on human agency in historical decision making, and its implications at the individual, group, national, and international levels (Banks, 2001). The HESI aims to provide teachers with information about the Holocaust as well as pedagogical knowledge to support their classroom instruction and lets teachers experience professional development as emancipatory praxis, so that they, in turn, can engage their students in meaningful learning.

Holocaust Education Summer Institute

This five-day HESI has been offered for the past eight years. A total of nine instructors, specialists in the fields of Holocaust history and critical pedagogy, participate in the program. On average, thirty teachers complete the program each year. Participants receive a range of materials including books, handouts, CD-ROMs of the Florida Holocaust curriculum manuals, and some Holocaust units. Below, we discuss the HESI's content and structure during a recent session to provide insight into the content and the pedagogy adopted each day in the weeklong curriculum.

Day One—Jewish Life in Europe before 1933. Day one emphasized the core goal of the program: moving from memory to action. The presenters underscored that it was important not only to remember the past but also to challenge feelings of helplessness and apathy by learning from the history to identify early signs of prejudice and prevent them promptly. Beginning with an evaluation of Jewish life in Europe before 1933, presenters discussed the crucial roles of religious texts and modern media in spreading anti-Semitism, highlighting the ways in which propaganda mobilized non-Jews to scapegoat and dehumanize the Jews. Out of this discussion teachers gained critical awareness of the need to help students learn how to deconstruct propaganda, to question the "unquestioned authority" of the holy scriptures, to ask critical questions, and to reinterpret history, in order to combat prejudice in today's society. To accomplish their pedagogical goal, facilitators (historians and theologians) used primary documents such as excerpts from the Hebrew Bible, as well as quotations from *Mein Kampf* and historical figures such as Tacitus and Martin Luther. In addition, they utilized visual examples of anti-Semitism, such as political cartoons, as important opportunities to advance student learning about the social dynamics of constructing prejudice.

Day Two—Germany and Nazi Persecution of the Jews. On the second day, the focus shifted to learning about Germany and the circumstances that helped the Nazis gain power (e.g., the economic concerns following World War I, the conditions of poor and middle-class people, the loss of faith in democratic government, and the ability of the Nazi leaders to identify with citizens' frustrations). To help teachers understand history from the perspective of German citizens, one facilitator engaged participants in a discussion of case studies representing German citizens' (professor, factory employees, lieutenant in the German army, grocery store owner, and attorney) frustrations and hardships during the Great Inflation after World War I, their perception of the ineffectiveness of Weimar Republic in restoring Germany's economy after the country's defeat in World War I, and their yearning for leadership that could restore Germany's pride. These case studies facilitated participants' comprehension of the political and economic climate of the time, the bases of Germans' choices under those circumstances, and especially their widespread loss of confidence in democracy as a viable form of government. Through constructs such as "bystander" or "upstander" roles among ordinary citizens, teachers explored their own responsibility as individuals and citizens to understand, critique, and raise their voices against unjust or discriminatory policies/actions.

These discussions prompted teachers to examine their experiences following the 9/11 tragedy, especially xenophobia. Teachers connected the bystander role with apathy regarding bullying, peer pressure, and social acceptance. They saw how discussions of history could become opportunities to engage their students in considering how decisions made out of fear and pressure could perpetuate injustice.

Day Three—Marginalization and Mass Murder. Day three drew on oral tradition to teach history, highlighting the power of personal story through the presentations of Holocaust survivors. Survivors shared their survival struggles during the Holocaust, including mass deportation, life in concentration camps, hard labor, mass murders, stories of escape, and the help of strangers. One survivor described how he and other Jewish children were saved by the people of one small French village, Le Chambon-Sur-Lignon, which served as an excellent example of moral courage and citizen dissent against the French government. The survivors' authentic voices helped teachers to identify with their struggles and to recognize the importance of understanding how hope, self-reliance, collective responsibilities, empathy, courage, and advocacy can contribute to the preservation of human life and the upholding of human dignity among oppressed people.

Day Four—Refuge, Resistance, and Rescue: European Jews, 1933–1948. The fourth day addressed Holocaust victims' efforts to resist Nazi laws, their attempts to seek refuge, and the reaction of the United States to the refugee crisis of 1933-1938 and again at the end of World War II. These issues were explored through discussion of primary documents relating to laws governing U.S. citizenship since the eighteenth century. Following an exploration of the changing definitions of citizenship in U.S. immigration laws, teachers viewed the documentary *America and the Holocaust*, which presented the ambivalence of the U.S. government toward Jews as well as the role of American anti-Semitism in the failure of the United States to use its power to save more victims. Not surprisingly, these discussions encouraged an analysis of current immigration policies in the context of human rights.

The documentary *The Long Way Home* not only presented the deplorable conditions of the concentration camps but also the Allies' poor understanding of the plight of Jewish victims and their failure to provide adequate humanitarian assistance to Holocaust survivors at war's end. After viewing these documentaries, teachers connected Jewish refugees' experiences to those of Iraqis and Somalis as well as to the victims of Hurricane Katrina. They discussed the urgency for individuals to take responsibility for victims of war and natural disasters rather than waiting for governments to act.

Day Five—Preventing Genocide Today: Challenges and Opportunities. The final day of the HESI expanded Holocaust study to the examination of genocide since 1945, bringing a more global perspective to study of the Holocaust. Facilitators emphasized participants' use of their newly acquired knowledge, not only to teach about genocide, but to teach "against" and to "prevent" genocide (Kennedy, 2005). Teachers viewed the documentary *Worse Than War*, which discussed genocides throughout the world, and what individuals and government could have done/should do to prevent them. Referring to the scene of Christian Serbs attacking Bosnian Muslims, one teacher noted that she could use the scene to challenge students to think about their beliefs vis-à-vis their actions. Teachers discussed that they could use the video in their classes to offer a worldwide perspective in the context of human rights.

Rethinking Social Studies Curriculum: Moving from Memory to Action

The HESI's explicit focus on the human rights aspects of the Holocaust underscores the value of teaching social studies from a humanized perspective. The HESI—despite its intense engagement with historical information—emphasized that the purpose of education is to go beyond knowing the facts about the past and moving towards action in the present. In this sense, the facts provide the foundation for developing more meaningful intellectual and potentially civic engagement. History was presented in terms of broad-based concepts such as bystander and upstander roles and the influence of deep-seated and unquestioned belief systems to make propaganda effective. Thus, despite teachers' limited knowledge about history, Holocaust education became much more than teachers learning about past events. It also encouraged them to prevent and intervene against prejudice and violence at micro and macro levels of human interaction. Furthermore, the HESI underscored the centrality of educators and the process of education to the development of critical awareness and social justice advocacy.

The human rights focus allowed for a sustained emphasis on human decision making through the historical events studied. These decisions were explored from multiple vantage points—Germans, Jews, rescuers, perpetrators, and bystanders. They explored decision making at the micro level of interpersonal relationships as well as at the macro level of groups, cities, and nations. Studying these decisions extended into discussions of genocide and prejudice around the world in times and places removed from Nazi Germany. The human story, whether through the survivor narratives or through the examination of marginalization in the context of political propaganda, compelled participants to recognize how one must actively engage with history in order to learn it (as opposed to passively accepting/memorizing fragmented and decontextualized information), and that effective learning is predicated on lessons being personally and socially meaningful.

The HESI as a Pedagogical Model for Social Studies Teachers

Grounded in the perspectives of critical pedagogy, the HESI also served as a model for instructional practice within schools. Its focus on Holocaust education was particularly apropos for demonstrating how social studies could/should facilitate critical awareness and democratic citizenship skills, as participants (and, in turn, their students) learned about the nature and function of propaganda; the role and use of religious beliefs to engender (or counter) prejudice; and specific examples of upstanding citizenship and resistance that required courage, persistence, and a commitment to human rights. The juxtaposition of this material, with emphasis on standardized tests and accountability regimes that teachers faced in their own schools, highlighted the struggle between emancipatory and oppressive potentialities of public education. The urgency of preventing bullying, intervening in genocides around the world, and developing a commitment to social justice while simultaneously addressing the pressures of standardized tests was acknowledged by all participants who discussed how they incorporated Holocaust education into their curriculum.

It does help teachers in Florida that Holocaust education is mandated by the state. However, there is limited support for teachers on how this should be incorporated into classroom practice. As an institute dedicated to professional development, the HESI provided both content knowledge and pedagogical knowledge (Borko, 2004; Eggen & Kauchak, 2006). Each session comprised specific historical information that offered participants the knowledge they lacked, while demonstrating how this content could be linked with broader concepts such as democratic citizenship through instructional practices that highlighted active and authentic learning. These instructional practices and sources of knowledge included the use of video documentaries (a medium that further enhanced the psychological power of the message), narratives of survivors, original historical documents, lectures from subject matter experts, and discussions with teachers about the classroom adaptations of the information and materials presented.

Professional Development in the Context of Globalization: Promises and Challenges

In highlighting the work of the HESI, this chapter attempts to address two educational imperatives: the need for social studies to be taught from a perspective that emphasizes concerns for human rights, and the need to provide teachers with professional development to enable them to do this work in contexts that require the narrowing of their curriculum in the face of standardized testing. Although it provides a concrete example of what can be done, this one-week program is often only a beginning for its participants.

While much was achieved in this short period, several challenges also surfaced. As might be suspected, teachers' ability to deliver this type of curriculum rests heavily on their own philosophical positions and belief systems. There was some evidence that while some participants were passionate about addressing the injustices faced by the Jews during the Holocaust, they were more reticent to see similar parallels of injustice in the contemporary experiences of marginalized groups (e.g., African Americans, Muslims, LGBT youth, refugees, and immigrants). While the HESI began to facilitate critical reflectiveness on participants' own assumptions, it revealed the need for ongoing reflexivity among teachers and the interconnection between teacher beliefs and attitudes and the successful implementation of emancipatory education.

Another challenge is the teachers' limited historical knowledge. Despite their willingness to learn, the need to address subject matter knowledge, engage with emotionally challenging topics, reflect on their own assumptions, and consider classroom applications made the experience exhausting as well as rewarding. This also reveals the complexity of the teaching process, a far cry from the teacher as technician model required in the movement towards teacher-proof curriculum. Furthermore, while participants were exposed to specific examples of how to integrate curriculum across diverse subject areas using nonfiction, fiction, documentaries, and primary documents, such examples are rare and underrepresented, even in teacher preparation programs.

At a time when the processes of globalization have eroded human rights around the world, the HESI represents a much-needed example of how social studies can be taught as a counter-globalization endeavor. While such an undertaking could be approached from numerous vantage points, the HESI represents one that addresses prejudice, conflict, and genocide. As an example of skillfully developed curriculum, the HESI integrates the intellectual, affective, and participatory domains through the provision of high levels of cognitive challenge, while also generating empathy for a diverse array of people and recognizing the need for civic action as an outcome of effective education. By highlighting divergent perspectives on how and why genocides occur, the participants were also exposed to decision making surrounding genocide; in turn, they could raise the question with their own students about what they would do if faced with a similar situation. The HESI serves as a reminder of how social studies can be taught (even in difficult times) and of the need for ongoing professional development among teachers to ensure that this occurs.

References

Abovitz, K. K., & Harnish, J. (2006). Contemporary discourses of citizenship. *Review of Educational Research*, 76(4), 653-690. doi:10.3102/00346543076004653

Apple, M. (2000). Between neoliberalism and neoconservatism: Education and conservatism in a global age. In N. Burbules & C. A. Torres (Eds.), *Globalization and education: A critical perspective* (pp. 59-77). New York: Routledge.

Au, W. (2009). Social studies, social justice: W(h)ither the social studies in high-stakes testing? *Teacher Education Quarterly*, 36(1), 43-58.

Banks, J. (2001). Teaching social studies for decision making and citizen action. In C. Grant & M. L. Gomez (Eds.), *Campus and classroom: Making schooling multicultural* (2nd ed., pp. 109-134). Upper Saddle River, NJ: Merrill/Prentice Hall.

Berger, J. (2003). Teaching history, teaching tolerance, Holocaust education in Houston. *The Public Historian*, 25(4), 125-131. doi:10.1525/tph.2003.25.4.125

Borko, H. (2004). Professional development and teacher learning: *Mapping the terrain*. *Educational Researcher*, 33(8), 3-15.

Bromley, P., & Russell, S. G. (2010). The Holocaust as history and human rights: A cross-national analysis of Holocaust education in social science textbooks, 1970-2008. *Prospects*, 40(1), 153-173. doi:10.1007/s11125-010-9139-5

Burroughs, S., Groce, E., & Webeck, M. L. (2005). Social studies education in the age of testing and accountability. *Educational Measurement: Issues and Practice*, 24(3), 13-20.

Eggen, P. D., & Kauchak, D. P. (2006). *Strategies and models for teachers: Teaching content and thinking skills*. Boston: Pearson.

Ekman, M. (2010). Exploring the relevance of Holocaust education for human rights education. *Prospects*, 40(1). doi:10.1007/s11125-010-9140-z

Feinstein, S. C. (2004). What are the results? Reflections on working in Holocaust education. In S. Totten, P. R. Bartrop, & S. L. Jacobs (Eds.), *Teaching about the Holocaust: Essays by college and university teachers* (pp. 51-63). Westport, CT: Praeger.

Freire, P. (1970/2000). *Pedagogy of the oppressed*, 30th anniversary edition. New York: Continuum.

Gatens, R., & Johnson, M. (2011). Holocaust, genocide and human rights education: Learning political competencies for 21st century citizenship. *International Journal of Social Studies*, 1(2), 35-47.

Holocaust Mandate, Fla. Stat. § 1003.42(f). (1994). Retrieved from http://flholocausteducationtaskforce.org/About/FloridaStatute

Kennedy, E. J. (2005). *Redefining genocide education*. Retrieved from http:// www.chgs.umn.edu/

Lindquist, D. H. (2010). Meeting a moral imperative: A rationale for teaching the Holocaust. *The Clearing House*, 84(1), 26-30. doi:10.1080/00098655.2010.496813

Loewen, J. W. (2008). *Lies my teacher told me: Everything your American history textbook got wrong*. New York: Touchstone.

Shiman, D. A., & Fernekes, W. R. (1999). The Holocaust, human rights, and democratic citizenship education. *The Social Studies*, 90(2), 53-62.

Sleeter, C. (2005). *Un-standardizing curriculum: Multicultural teaching in standards-based classrooms*. New York: Teachers College Press.

Stromquist, N. P. (2002), *Education in a globalized world: The connectivity of economic power, technology and knowledge*. New York: Rowman & Littlefield.

Totten, S. (1997). A note: Why teach about the Holocaust? *Canadian Social Studies*, 31, 176-177.

Zinn, H. (2007). Why students should study history. In W. Au, B. Bigelow, & S. Karp (Eds.), *Rethinking our classrooms* (Vol. 1, pp. 179-185). Milwaukee, WI: Rethinking Schools.

CHAPTER NINETEEN

Preparing Teachers for Global Consciousness in the Age of Globalization

Lydiah Nganga

> Preparing teachers for global consciousness is critical because schools are increasingly diverse. To ensure that educators have the knowledge and skills needed in this globalized era, it is important for teacher education programs to have course content that examines global perspectives. This chapter explores pre-service teachers' perceptions of global citizenship, discusses instructional activities used in a social studies methods course to help pre-service teachers to develop skills critical to cultural appreciation, and, finally, it examines the application of the knowledge gained to a social studies curriculum.

The ongoing globalization phenomenon is not limited to global economies. Rather, other areas of life are affected as well. For example, it appears that students are seeking college education at rates never before seen, perhaps because of the pressure of globalization. In essence, "keeping pace in a world of rapid technological change puts a premium on being flexible and prepared to respond to unknown situations and challenges" (McPherson & Schapiro, n.d., p. 4). To McPherson and Schapiro, globalization requires higher education curricula to factor international contexts. Indeed, in the current age of rapid globalization, institutions of higher learning are at the crossroads of many cultures due to increased global migration and mobility. As a result, they must be ready to teach skills and impart knowledge essential to global consciousness. In addition, due to increasing interconnections between nations, and as nations continue to face global issues, there is a need to broaden existing higher education curricula to include democratic values and cultural pluralism within the context of diversity and global interdependence (Kambutu & Nganga, 2008; Nganga, 2009; Nganga & Kambutu, 2012).

Educating college students for global citizenship should be a priority. While many skills are needed, an education that develops respect and appreciation of all cultural groups is essential (Ameny-Dixon, n.d; Marchetto, 2010). Thus, colleges of education in the United States are increasingly utilizing a variety of instructional strategies in order to prepare globally minded teachers who are able to work well with international and diverse communities (Alfaro, 2008).

Theoretical Perspectives

Due to globalization, all teacher education programs should implement curricula that teach essential global skills. Reflecting on the value of global curricula, Bleicher and Kirkwood-Tucker (2004) indicated that infusing a global perspective in the teacher education curriculum is necessary because such a curriculum helps pre-service teachers to gain a better understanding of multiple perspectives while promoting an appreciation of diverse viewpoints. Preparing teachers for a global consciousness means helping them to implement "content and global competencies which incorporate indispensable information, skills, and attitudes about the world" (McCabe, 1997, p. 41). In addition, developing a global consciousness creates essential space to facilitate the understanding of human circumstances while exhibiting intellectual curiosity about the world that transcends national borders (Merryfield, 1997; Nganga, 2009; Nganga & Kambutu, 2012).

Global consciousness helps individuals to examine their values and assumptions and to realize that one's views and values are not necessarily universal. Instead, there exists a diversity of ideas and practices across cultures. Hanvey (1982) reckons that such a realization is incorporated into new capacities of understanding the world and is an important foundation for global consciousness. In his model, Hanvey (1982, p. 162) proposed a conceptual framework for global education that includes the following five dimensions: 1) Perspective consciousness—understanding of multiple perspectives; 2) state of the planet awareness—an understanding of world conditions and emerging trends; 3) cross-cultural awareness—understanding similarities and differences between different cultures; 4) knowledge of global dynamics—understanding global interdependence; and 5) awareness of human choices—developing an awareness of the challenges of individual, local, and national choices. In addition, topics that explore economic and power differentials within and across nations should be addressed. In this chapter, I explore a variety of instructional strategies that I apply in a social studies methods course for pre-service teachers to help my students gain a better understanding of the world.

Social Studies and Multiple Perspectives

Because teachers are charged with the responsibility of educating future leaders, pre-service educators should encounter multiple opportunities to help them acquire essential knowledge, skills, and dispositions. Indeed, in the current era of globalization, education for global consciousness is needed (Zarrillo, 2012). An education for global consciousness involves imaginative exploration and discovery of global perspectives. In addition, a curriculum that shows how people are linked in a global economy and political, ecological, and technological systems is essential. Equally important is a curriculum that focuses on world issues and solutions, human interactions and culture, universal human rights, and the role of international organizations.

Given the existing need for an education for global consciousness, I developed instructional units that exposed my pre-service teachers to global education. The units I designed had several learning opportunities intended to help learners to confront stereotypes and misconceptions about other cultures and also to examine issues of dominance and social justice. By incorporating a curriculum for global consciousness, my primary goal was to help my students to make connections, comprehend, and develop an appreciation of other cultures. Therefore, at the beginning of the semester, I planned an activity that required them to define global education and to explain, based on their perceptions, characteristics of a globally conscious educator. Their definitions were very narrow. Generally, they defined global education as the awareness of where we live. A few students provided broader definitions such as the study of the geography of the world and learning about different cultures. But a majority of my students showed ethnocentric tendencies, that is, judging other cultures negatively using their own culture as the norm. By the end of the semester, however, their responses to the same question had changed.

After experiencing a semester-long curriculum for multiple perspectives, I again asked my students to define globalization. Their responses showed that they had gained a better understanding of global education. For example, one student said global education involved "teaching through a global perspective. It is acknowledging that we are citizens of the world and what we do affects so many more than what we see in front of us. It is acknowledging and embracing multiple perspectives" (class discussions, Fall 2011). Other students appeared to develop critical awareness as well. To that end, one student defined global education as "opening people's eyes and minds to the realities of a globalizing world and helping them to bring about a world of greater justice, equality, and rights for every individual." Other learners viewed global education as "learning about the world and not just about the area that one may live in. It opens a door to every issue in every aspect." Elaborating on this view, a student provided the following description:

> We are all global citizens. With the advent of globalization and free trade agreements as well as advances in technology, the world has become a much smaller place. A person can travel halfway around the world in a day. Such a journey would have taken months just a century ago. There is almost no place on the planet where you cannot get cell phone reception and be in contact with someone half a world away. With

these advances in technology, it is more apparent than ever that we all share this amazing planet and our actions are all interconnected (Course reflections, Fall 2011)

While my students' perceptions at the beginning of semester were apparently shallow, post-teaching reflections showed that they had developed a deeper understanding of global education, perhaps due to experiences with a variety of instructional strategies.

Instructional Strategies

One of the most helpful instructional strategies that I used involved the study of cultural similarities and differences. Although potentially challenging, when teaching is structured thematically, learners are able to make connections across broad reaches of time and space (Levstick & Barton, 2011). Helping students to make connections across time, for example, from past to present, shows them the link between past and present. To achieve this goal, I structured my instruction around thematic units that supported a holistic study of global issues. Again, one unit focused on the similarities and differences between foreign nations and the United States. To ensure critical learning, Martorella (1985) cautioned against planning a curriculum that focuses on the colorful without drawing deeper meaning. Other scholars, for example, Nganga (2006), Merryfield and Wilson (2005), and Derman-Sparks and Edwards (2010), were against teaching approaches that highlight exotic rituals and foods only because they cause learners to develop impressions that other countries are strange and weird. Instead, a curriculum that helps learners to develop understanding, appreciation, and sensitivity to differences is ideal. Education for cultural sensitivity helps learners to develop tolerance, respect for diversity, open-mindedness, and a willingness to consider divergent arguments and solutions. Meanwhile, instructional vignettes are equally effective, especially because when planned well, they allow students access to prior knowledge, misconceptions, and attitudes towards differences. As a result, I utilized multiple vignettes to teach to global differences.

Vignette Activity. To help my students access their existing knowledge, misconceptions, and attitudes relative to global differences, I planned several vignettes that focused on international cultural experiences. As an example, I have provided a vignette below that my students studied before responding in writing to corresponding divergent questions.

Preparing Teachers for Global Consciousness

> Vignette: Global education requires teachers to be competent both in national and international issues. Indeed, all teachers in a globalized era should have intercultural awareness. So, as a fifth-grade educator, you have been awarded an international travel grant to take your students to Kenya. This is a great opportunity for you and your students to learn about new cultures. After writing a letter to your students' parents to inform them of the impending exciting cultural experience, all of them are happy and want to know how to prepare their children for the trip. However, because this trip will be your first to Kenya, you are not sure how to respond to parents' questions. As a starting point, you need to respond to the following questions:
> 1. How would you get ready for this international experience?
> 2. What problems would you anticipate?
> 3. Why do you anticipate the problems you do?

Although my students responded to the vignette's questions divergently, there was evidence that, while they had effective research strategies, they generally lacked essential cultural information. Consider, for example, the following response that shows a lack of existing knowledge about Kenya:

> It would be important to seek all the information I can find on Kenya. I possibly would need to research the customs, food, and information that would help me and my students not to be intentionally disrespectful. I would research the climate. I would let students watch videos, read books, and do an Internet search. (Class reflection, Fall 2011)

The issue of safety or lack thereof was a concern among many students. A majority of them reported that it might not be safe for Americans to travel to Kenya. As a result, one pre-service teacher indicated, "I would need to know about safety for Americans. Many foreign countries do not like Americans. I would also find out how parents and relatives would reach their children while in Kenya, including contact persons in Kenya and arrange for an education guide." Meanwhile, other responses showed many misconceptions. For example, there was a general wrong perception about the "lack of clean water and toilets" in Kenya. Other students viewed Kenya in the context of "wild animals roaming everywhere and naked tribal people." The threat of "catching strange diseases such as HIV/AIDs" in a country "without medical services" was an additional misconception. Meanwhile, my students were concerned about the possibility of "not being able to communicate with people back home (U.S.) due to lack of telephone services in Kenya."

Based on pre-service teachers' responses, it was clear that they lacked important global competencies that are critical to educating for global citizenship. In their responses, for example, they appeared to fear foreign cultures. But most surprising was an apparent resistance to learning about other cultures. In addition, other problems, such as lack of flexibility to accommodate other cultures, an inability to deal with the possible stress of meeting new cultures, and an unwillingness to relate to people of other cultures, were evident as is apparent in the excerpts provided below from two students.

> I really do not know why we have to learn this. It is a waste of time. Schools are not even teaching to diversity. I plan to teach early childhood; I do not think I will take my students overseas, so this is pointless.

> My personal biases could hinder my full participation in cultural experiences. I am not a risk-taker mostly because I have never lived alone. I have always been around people that I know. I think taking students for an overseas trip requires one to be comfortable at taking risks. I would anticipate problems adjusting to the environment, allergies, lack of clean water and foods.

Notwithstanding the apparent misconceptions, subsequent instructional processes, including library research of a foreign country of their choosing, helped my students to develop informed conceptions.

Studying a Foreign Nation

The purpose of this project was to help my learners to develop a deep comprehension and appreciation of other cultures. In this activity, learners studied a country of their choice. Using primary and secondary resources, they wrote a research paper that not only examined the similarities and differences between the studied country and the United States but also looked at economic disparities, impact of cultural differences, and systems of domination. I encouraged learners to take different perspectives relative to the interpretation of sensitive issues, and most important, to engage in transformational perspectives. In support of the transformational approach to learning, Banks (1997, p. 237) expressed the importance of viewing "concepts, issues, themes, and problems from multiple perspectives." In addition to promoting empathy, having transformational perspectives creates an ideal space for critical analysis of complex global realities. Meanwhile, the following rubric guided the study of a foreign country:

- Researching and writing a five-page paper that included the studied country's geographic location, cultures, people, and languages.
- In-depth comparison between the studied country's major institutions and those in the United States.
- Applications; that is, a discussion of how the student might use the knowledge gained as a social studies teacher.

- Presentation: prepare a brief oral presentation to the whole class. This culminating activity required learners to share findings, reflect, and personalize in order to explore existing beliefs and teaching practices relative to global issues (Johnson & Ochoa, 1993). A sharing of artifacts and foods was also expected.

The assigned research activity helped my students to develop a deeper understanding about global issues. In addition to helping them confront their misconceptions, they were able to broaden their views and to see that "each nation has its own diversity of lifestyles, range in socio-economic levels, mix of economic activities, and set of differences between rural areas and big cities" (Brown & Carroll, 2008, p. 12). The following quotes by two students reflect obvious growth:

> I really enjoyed the study of my country. I learned so much about the cultural differences and similarities. It is amazing how people are the same. What we consider as lacking is seen as wealth by other nations.

> I really loved the unit. I enjoyed researching about a country that I was interested in, and really liked the kinesthetic mapmaking with dough. The food gave a visual presence of the country we each researched and of course a taste of the country. The whole project and peer presentations helped me to see the differences and similarities in the countries, the religion, the history, and much more. For example, I did not know that most countries required primary education (0–16 years). I also thought that some countries were worse off than they really are. This unit, therefore, opened my eyes, and served to educate my ignorance.

Exploring Multiple Perspectives

My class provided multiple opportunities for learners to experience a variety of perspectives. Due to isolation from ethnic communities, my students, who are mostly white, middle class, and residing in a mostly rural state, lacked essential diversity awareness. Therefore, I selected learning materials and teaching strategies that challenged their notions of culture, justice, heroes, and heroines. Specifically, I applied a discourse of multiple perspectives, including rethinking Columbus, slavery, colonization, and neoliberalism in the context of the ongoing globalization efforts.

Rethinking Columbus. Columbus played a key role in spreading globalization. Indeed, many people, indigenous nations especially, are still affected by Columbus's activities. Thus, teaching about Columbus requires a deep examination of his exploratory expeditions, both the good and the bad. In the absence of a critical examination, Columbus's work will continue to rationalize many injustices against indigenous people. To that end, Bigelow and Peterson (1998) reported that "children's biographies of Christopher Columbus function as primers on racism and colonialism" (p. 47). In addition, these biographies perpetuate the notion that it is acceptable to dominate "other" people. Consequently, the Columbus myth of "discovering America" is very popular in history classes. According to Bigelow and Peterson, the myth stops

> children from developing democratic, multicultural, and anti-racist attitudes. Children's biographies of Columbus depict the journey to the New World as a great adventure, led by probably the greatest sailor of the time. It's the story of courage and superhuman tenacity. Columbus is brave, smart and determined. But behind this romantic portrayal is a gruesome realty. For Columbus, land was real-estate and it didn't matter that other people were already living there; if he "discovered" it, he took it. If he needed guides or translators he kidnapped them. If his men wanted women, he captured sex slaves; if the indigenous people resisted, he countered with vicious attack dogs, hangings, and mutilations. (Bigelow & Peterson, 1998, p. 47)

In my course, therefore, we counter various myths using Bigelow and Peterson's (1998) ideas. Some of their ideas that we study include the following: "we have no reason to celebrate an invasion" (pp. 12–13); "discovering Columbus: re-reading the past" (pp. 17–21); "sugar and slavery" (pp. 22–23); "Columbus and native issues in the elementary classroom" (pp. 35–41); and "the people vs. Columbus" (pp. 85–88). Meanwhile, we use other reading resources such as Jane Yolen's *Encounter*, a children's book that tells additional stories about Columbus from the American Indians' viewpoint. These reading resources provide conflicting points of view, thus helping my pre-service teachers to develop mental flexibility while gaining new perspectives on previous learning. I also use dialogue to extend their thinking.

Dialogues allowed my students opportunities to vocalize any inconsistencies they had between previously held notions about Columbus and the information encountered in my course. The goal was to encourage learners to change their conceptions as needed and to also think about ways to counter Columbus's myths in their own classrooms. To connect this learning activity to current global issues such as immigration, trade, and human rights, I asked my students to respond to the following questions from McNulty, Davies, and Maddox's (2010, p. 22) work:

1. What impact did trade have on exploration?
2. What influences trade today?
3. Describe the influences of governing bodies of each party during Columbus's exploration

4. What impact did physical geography have on Columbus's exploration?
5. What would you consider to be modern-day exploration?

After discussing these questions, a dialogue on why Native Americans might find "no reason to celebrate an invasion" ensued. This discussion helped the students to understand the complexity of current global issues, including immigration and economic exploitation of poor countries by the wealthy nations of the West. In addition, students were able to see the complex nature of Columbus's activities even though "too often, this story is posed as a romantic myth, and uncomfortable facts about Columbus are eliminated" (Bigelow & Peterson, p. 13). However, educators should understand that teaching the unpleasant truths about history does not take away history's positive side. From the unit they completed, my students learned that global issues are complex, there are many sides to a story, and it is therefore important to ask whose story it was and why it was told the way it was. They also learned the importance of learning and teaching activities that promote mental flexibility while taking a stand to change misconceptions.

Conclusions and Implications for Social Studies Educators

This chapter has explored the phenomenon of globalization and its impact on teacher education. The challenges presented by globalization require teachers who are ready to educate for a global consciousness because "children's learning must be rooted in the ethical imperatives of global citizenship, which require new ways of finding information, making connections, and listening to diverse voices from the world in response to pressing global challenges" (Heilman, 2008, p. 32). To teach for global consciousness requires teachers to have an understanding of cultural universals and cultural diversity and an appreciation of multiple perspectives. Teachers need to have knowledge of global systems, interdependence, and interconnections. In the current reality of globalization, all educators should be prepared to teach about global issues because they "cannot prepare youth for their future without an appreciation and knowledge of the world" (Merryfield, 1991, pp. 18–19). It is equally critical to acquire pedagogical skills essential to teaching for a global perspective.

In an attempt to teach for a global perspective, I challenged my students by asking them to study topics that did not align with previously held notions. In the end, however, my students felt that such learning was necessary because it helped them to understand that history is complex and developing solutions to world problems requires people with a global consciousness. The following response relative to Columbus's myths captures the benefit of teaching for global perspective rather well:

> We feel pretty annoyed that we were taught wrong. It makes us feel ignorant but by the same token, we are happy we had this opportunity to re-learn history. It is every teacher's responsibility to put real information in learners' minds because they learn what is taught. We felt lied to and this is gross injustice in education. What is worse is that this mis-education is still being taught. Someone needs to do something to counter the misconceptions that teachers hold and are being transferred to their students.

Helpful in broadening my students' global perspective was the study of a foreign country. To that end, one learner reported that

> it is surprising to learn that some of these countries have better health systems than the United States. In this project, I saw the many cultures of the world represented in our country and elsewhere. This project was such a tool for creating cultural awareness. There are many commonalities between countries of the world, many similar forms of government, education, and other things. It was amazing to learn that women generally live longer than men in almost all countries. There were similarities also in forms of religions.

At the beginning of the semester, my students were uncomfortable with unfamiliar cultures. But after experiencing challenging instructional activities relative to the power of cultural understanding and appreciation in a globalized era, they developed a desire to teach for multiple perspectives in order to help their students develop empathy, cultural understanding, and appreciation. The use of instructional strategies that help learners to engage in multiple ways, such as vignettes, the study of similarities and differences between nations, and exposure to literature that helps learners to develop mental flexibility and perspectives, is essential in the process of promoting global consciousness.

When used to emphasize global knowledge, connectivity, and understanding of human interactions, teaching for multiple perspectives helps learners to develop a deeper understanding of global issues. Nevertheless, because teaching for multiple perspectives requires learners to examine and modify existing erroneous perceptions and assumptions, educators should always prepare a foundation essential to confronting stereotypical thinking and misconceptions. In the absence of an ideal foundation, some learners might be emotionally threatened, thus causing them to disengage. But because social studies educators have a responsibility to teach for effective global citizenship, they should always be prepared to address any unexpected challenges as they teach for creative thinking and flexibility.

References

Alfaro, C. (2008). Global student teaching experiences: Stories bridging cultural and intercultural difference. *Multicultural Education*, 15(4).

Ameny-Dixon, G. M. (n.d). *Why multicultural education is more important in higher education now than ever: A global perspective.* Retrieved from http://www.nationalforum.com/ Electronic%20Journal%20Volums/Ameny-Dixon%2C%20G

Banks, J. A. (1997). Approaches to multicultural curriculum reform. In J. A. Banks & C. A. McGee Banks (Eds.), *Multicultural education: Issues and perspectives* (3rd ed., pp. 229-250). Boston: Allyn & Bacon.

Bigelow, B., & Peterson, B. (1998). *Rethinking Columbus: The next 500 years. Resource for teaching about the impact of the arrival of Columbus in the Americas.* Milwaukee, WI: Rethinking Schools.

Bleicher, R. E., & Kirkwood-Tucker, T. F. (2004). Integrating science and social studies teaching methods with a global perspective for elementary preservice teachers. *Curriculum and Teaching Dialogue*, 6(2), 115-124.

Brown, B. B., & Carroll, A. (2008). Beyond wildlife: Teaching about African stereotypes. *Social Studies and the Young Leaner*, 20(4), 12-17.

Derman-Sparks, L., & Edwards, J. O. (2010). *Anti-bias education for young children and ourselves.* Washington, DC: National Association for the Education of Young Children.

Hanvey, R. (1976). *An attainable global perspective.* Washington, DC: American Forum for Global Education.

Hanvey, R. (1982). An attainable global perspective. *Theory into Practice*, 21(3), 167.

Heilman, E. E. (2008). Including voices from the world through global citizenship education. *Social Studies and the Young Learner*, 20(4), 30-32.

Hicks, D. (2003). Thirty years of global education: A reminder of key principles and precedents. *Educational Review*, 55(3), 265-275.

Johnson, M., & Ochoa, A. (1993). Teacher education for global perspectives: A research agenda. *Theory into Practice*, 32(1), 64-68.

Kambutu, J., & Nganga, L. (2008). In these uncertain times: Educators build cultural awareness through planned international experiences. *Teaching and Teacher Education*, 24(4), 939-951.

Kirkwood, T. F. (2001). Our global age requires global education: Clarifying definitional ambiguities. *The Social Studies*, 92(1), 10-15.

Levstick, L. S., & Barton, K. C. (2011). *Doing history: Investigating with children in elementary and middle school* (4th ed.). New York: Routledge.

Marchetto, A. (2010, June). *Higher education in the global context.* Paper presented at the ACCU (Association of Catholic Colleges and Universities) seminar. Available from Zenit Web site: http://www.zenit.org/article-29718

Martorella, P. H. (1985). *Elementary social studies: Developing reflective, competent, and concerned citizens.* Boston: Little, Brown.

McCabe, L. T. (1997). Global perspective development. *Education*, 118 (1), 41-46.

McFadden, J., Merryfield, M., & Barron, K. R. (1997). *Multicultural and global education: Guidelines for programs in teacher education.* Washington, DC: American Association of Colleges for Teacher Education.

McNulty, C.P., Davies, M., Maddox, M. (2010). Living in the global village: Strategies for teaching mental flexibility. *Social Studies and the Young Learner*, 23 (2), 21-24.

McPherson, M., & Schapiro, M. O. (n.d). *Global issues in higher education: What colleges should know*. Available from Educause Web site: http://net.educause.edu/r/library/pdf/

Merryfield, M. (1997). A framework for teacher education in global perspective. In M. M. Merryfield, E. Jarchow, & S. Pickert (Eds.), *Preparing teachers to teach global perspectives: A handbook for teacher educators* (pp. 1-24). Thousand Oaks, CA: Corwin Press.

Merryfield, M., & Wilson, M. (2005). *Social studies and the world: Teaching global perspectives*. Washington, DC: National Council for the Social Studies.

Nganga, L. (2006). Engaging in multicultural curriculum dialogues: Moving from a tourist approach to an anti-bias curriculum. *Journal of Early Childhood and Family Review*, 13, 31-43.

Nganga, L. (2009). Global and cultural education prepares preservice teachers to work in rural public schools. Teaching for social change in the 21st century. *Journal of Education Research*, 3(1/2), 149-160.

Nganga, L. & Kambutu, J. (2012).Broadening social studies curricula: Integrating global education in a teacher education program. In Russell, W.B. (Ed.), *Contemporary Social Studies: An Essential Reader (Teaching and Learning Social Studies)* pp. 527-546. U.S.A. Information Age Publishing

Zarrillo, J. (2012). *Teaching elementary social studies: Principles and applications* (4th ed.). Boston: Pearson.

CHAPTER TWENTY

Creative Pedagogies in Integrating Global Awareness in Secondary Social Studies Curricula in Teacher Education Programs and Schools

Toni Fuss Kirkwood-Tucker

> The teacher is the ultimate gatekeeper of what is being taught in the classroom.
> S. Thornton, 2005

Social studies teachers play a key role in preparing students for effective citizenship in a global age. The challenge of this responsibility is reflected in the following selection from *The Wind in the Willows* (Grahame, 1908), which provides a wonderful springboard when you first introduce the concept of global awareness to your students:

> "So-this-is-a-River." The Mole had never seen a river before. "The River," corrected the Rat. "But what lies over there?" asked the Mole . . ."That? Oh, that's just the Wild Wood," said the Rat shortly, "we don't go there very much, we river-bankers." "Aren't they—aren't they very nice people in there?" asked the Mole a trifle nervously. "W-e-ll, let me see. The squirrels are all right. And the rabbits—some of 'em; the rabbits are a mixed lot. And then there's the badger, of course . . . nobody interferes with him. They'd better not," he added significantly. "Well, of course—there are others, weasels and stoats and foxes and so on. They're all right in a way, I'm very good friends with them, pass the time of day when we meet and all that—but . . . well, you can't really trust them and that's the fact." "And beyond the Wild Wood again?" the Mole asked, "where it's all blue and dim . . . ?" "Beyond the Wild Wood comes the Wide World," said the Rat. "And something that doesn't matter either to you or to me. I've never been there, and I'm never going, nor you either if you've got any sense at all. Don't refer to it again, please. Now then, here's our backwater at last where we're going to have lunch."

Is it our role to help students to cross the river, to enter the wild wood and explore the wide world to promote cross-cultural understanding and interconnectedness among people and nations in the twenty-first century and beyond?

Introduction

In the age of globalization, classroom teachers are expected to invent creative strategies to infuse a global awareness in social studies curricula and instruction required by state departments of education and school districts. Global educators are responding to the call by professional organizations, business communities, and national commissions to add a global dimension to teacher education programs in order to help teacher candidates[1] acquire explicit knowledge, skills, and dispositions to teach from a global perspective (Kirkwood-Tucker, 2009; Merryfield, 1997, 2009; Quezada & Cordeiro, 2007). Yet global educators are still concerned about the lack of global content knowledge and geography skills of their undergraduate and graduate teacher candidates. Equally concerning is the absence of a systematic infusion of global awareness in requisite curricula in school systems across the United States by classroom teachers (B. Cruz, interview, March 10, 2012).

Globalization is the primary reason that global awareness education is a more viable concept today than it has ever been (Heater, 2002). According to Myers (2006), despite the reality that "scholarship on globalization suggests new forms of democratic citizenship and politics, the U.S. educational system remains resistant to global awareness teaching in the curriculum to favor national identity and patriotism over learning about the world" (p. 1). Most often, however, teacher candidates in teacher education programs and teachers in the trenches, with little exception, have not been trained how to integrate a global awareness in curriculum and instruction.

Possessing a global awareness facilitates the building of bridges across cultural boundaries to communicate and collaborate with those whose attitudes, values, knowledge, and ways of life significantly differ from their own (Cushner & Brennan, 2007; Kirkwood-Tucker, 2001a, 2002, 2003; Kirkwood-Tucker & Benton, 2002). This chapter offers creative pedagogies for teacher educators in teacher education programs and classroom teachers in schools how to effectively teach about the world and its people despite extant state and district curricular requirements. My proposed pedagogies will assist future teachers as well as classroom teachers to acquire an expanded knowledge about the historical, socioeconomic, political, and geographic interconnectedness of the world so that they will possess the competencies needed to integrate a global awareness in their teaching. Although practicing teachers often feel overwhelmed when it comes to adding anything new to the teaching process, these flexible strategies can be effectively integrated in daily, weekly, or monthly deliberate instruction to prepare our youth for competent, cooperative, and humanistic citizenship in the global age.

Integrating a Global Awareness into Curricula and Instruction

The following section of this chapter proposes five distinct pedagogies for how teachers can bring a global awareness to required curricula and instruction by 1) postholing global issues; 2) infusing comparative approaches in history, economics, and government courses; 3) integrating contemporary issues into traditional content; 4) teaching current events from a global perspective; and 5) simulating the United Nations. I describe, explain, and offer specific examples of how to teach from a global perspective.

In selecting examples for classroom teachers and teacher candidates, I have chosen history, economics, and political science courses because they represent the core courses in secondary social studies programs in schools, known as U.S. History, World History, Economics, and Government. These courses are the most frequently taught and, most likely, will be the very courses our graduates from teacher education programs will teach in their future classrooms.

Postholing Global Issues

My first suggested pedagogy in bringing a global awareness to the required social studies curricula in schools involves postholing lessons of a global issue at an opportune time during the instructional process. The "postholing" pedagogy facilitates an in-depth examination of a topic deemed important by the teacher by including one or more comprehensive lessons into the existing curriculum framework. This requires speeding up or condensing the required curriculum time line with a summarizing film or lecture (surely to the exclusion of battle simulations) to free up class time to make room (posthole) for lessons or an entire unit of instruction taught from a global perspective on a monthly basis throughout the school year.

In schools, the postholing of a global issue can also be implemented at the end of a unit; before vacations; or at the very beginning or completion of the school year when administrative disruptions and restlessness of students pose disciplinary problems. From my experiences in Florida's public schools, Advanced Placement teachers often teach about the world after students have completed the notorious Advanced Placement examination. Since teachers are the ultimate gatekeepers of what is being taught in the classroom (Thornton, 2005), they clearly decide whether to teach about one topic such as poverty and its ramifications or several lessons on different global issues. As instructional gatekeepers, teachers decide the number of lessons they are able to posthole into the mandated curriculum, because the lessons must be aimed at a specific group of students and be tailored to their particular interests and needs in the embellishment of the extant curriculum.

In teacher education programs, teacher educators can require future teachers to develop several lessons or a unit plan on a global issue(s) as part of the assignments in methods courses which teacher candidates, upon landing a teaching position, can posthole into their future social studies classes. At Florida State University, I require a unit plan of five lessons in the *Developing a Global Perspective* course required of all undergraduate and graduate students. After grading and consequent revisions, I require the placement of these lessons in teacher candidates' portfolios.

Global issues can include child labor; child soldiers; gender discrimination; global warming and its effects on climate and living; HIV/AIDS; human trafficking; infant mortality rate in industrialized and nonindustrialized nations; lack of primary education; plight of farmworkers worldwide; effects of population growth; poverty and malnutrition; sexual exploitation of women and men and minors; slave labor; the widening economic gap between industrialized and nonindustrialized nations; or other issues confronting the world and its people deemed important by the teacher.

Human rights violations. One example of a "postholing" pedagogy is the importance of teaching a lesson about human rights violations occurring in any part of the world, including the United States. Using the Universal Declaration of Human Rights (UDHR) (United Nations, 1948) and the Convention of the Rights of the Child (CRC) (United Nations, 1990), the two seminal documents created by the United Nations, a group activity can include downloading case studies from the Web about child labor and adult slave labor (or sexual exploitation of women and girls), and having student teams identify the human rights violated according to these documents. Drawing two charts on the board outlining the rights of the UDHR and the CRC, student teams check off which human rights were violated in their assigned case study. This activity challenges students to discuss the overarching nature of human rights violations, expanding their global awareness about pervasive injustices in today's world.

Infusing Comparative Approaches in History, Economics, and Government Courses

A second pedagogy in bringing a global awareness to requisite social studies courses in schools involves comparison of a historical event required to be taught in the curriculum with a similar event that occurred in the recent past or is occurring in the present in another part of the world.

History. For example, in teaching U.S. history, the teacher draws a comparison between the American Revolution and a revolution in more recent times such as the Cuban Revolution or, in contemporary times, the 2011 Libyan Revolution.[2] Students compare and contrast commonalities and differences of these revolutions and identify the overarching reasons for their outbreak. This method widens the teaching of U.S. history from a nation-centered or ethnic-centered perspective to a global perspective of similar events occurring in nations around the world. Other historical comparative approaches include the American independence movement compared to the independence movement in Ghana or India; the U.S. Civil War compared to the civil war in Sudan and the new nation of South Sudan; the U.S. civil rights movement compared to the apartheid movement in South Africa; the U.S. civil rights movement compared to the Palestinian independence movement from Israel; and African American uprisings/demonstrations/sit-ins compared to contemporary uprisings in Bahrain, Egypt, Libya, Syria, Tunisia, and Yemen.

Economics. In economics classes, comparative examples include the U.S. rise as an economic power compared to the emerging economic powerhouse of Brazil; the changing role of the United States as the world's economic superpower compared to the recent changes in BRIC nations (Brazil, Russia, India, China); U.S. child labor during the Industrial Revolution compared to child labor in Bangladesh today; U.S. economic interconnectedness with Canada, China, and Brazil compared to their interdependence with the United States; U.S. migrant laborers' economic struggle compared to Chinese/South Korean migrant laborers in the Middle East.

Government. In government classes, comparative examples can include the U.S. Constitution compared to the Constitution of South Africa; the U.S. judicial system compared to the judicial system of the People's Republic of China; the U.S. penal system compared to the penal system of Iran or Mexico; U.S. immigration policies compared to those of the European Union (EU); U.S. Medicare/Medicaid policies compared with national health insurance in the Federal Republic of Germany.

It is important that the teacher makes these comparisons without going into too much depth, as time constraints are teachers' greatest frustrations. By simply referring to the compared event, students are exposed to global awareness by being introduced to the concept of connectedness between the United

States and other nations. The comparative approach does, however, require a deeper knowledge base in economics, history, and political science on the part of the teacher.

Integrating Contemporary Issues into Traditional Content

A third pedagogy in bringing a global awareness to required social studies courses is the integration of contemporary global conditions that exist in the community of nations as a result of past historical oppression of a people.

World history. In teaching world history, for example, when addressing the cause-and-effect relationship of colonialism, the teacher can demonstrate the interconnectedness of the historical event of the past and the present-day condition as a result of that past oppression. History has shown us that many formerly colonized countries have difficulty overcoming vestiges of colonialist practices, as evidenced in Latin American and African countries. Indonesia is also a case in point. As a direct or indirect result of Dutch colonialism, a large number of Indonesians today are suffering from hunger and malnutrition. Many of their children today are born to undernourished mothers in deep poverty, resulting in at-risk student populations that suffer from emotional, mental, and physical ailments. After giving this example, the teacher challenges students: Does colonialism in some countries have lasting effects on its population? Why? In your opinion, why might this be the case? How can an earlier oppressed culture recover from colonialism? What will be the future of these impaired children? Do these conditions have an impact on their country? Its future? Can this issue be remedied? Does the former colonial power have a moral responsibility to assist the Indonesian government in alleviating poverty? Who can show us Indonesia on the world map? What is its capital?

Economics. The economics teacher, when introducing the concepts of supply and demand and diminishing returns in the global marketplace, can provide alarming insight into the use of forced labor by transnational corporations, resulting in excessive corporate profits. When teaching about immigration, the teacher can introduce the concept of remittance and give examples of Mexican, Haitian, and other foreign laborers who work in the United States. These workers send home large percentages of their wages, which may mean the very survival of their extended families and/or entire communities.

Government. The government teacher can compare and contrast the social rights of citizens in Communist nations with those of the poor in the United States, many of whom suffer from higher unemployment, underemployment, lack of education, and economic deprivation than the mainstream population. The teacher may bring the concept of "being my brother/sister's keeper" into a dynamic discussion evaluating whether the role of government assistance to the underprivileged is warranted.

Geography. The geography teacher can compare and contrast prevailing global issues in particular regions of the world or local communities, such as infant mortality rate among ethnic groups. For example, in Florida, the average white infant mortality rate from 2007 to 2009 was 5.2 per 1,000 live births, whereas the African American infant mortality rate per 1,000 live births was 13 (Ames, 2011). Identifying a local social issue and comparing its similarities or differences with a social issue in another society facilitates cross-cultural understanding that can lead to empathy and transpection (Hanvey, 1976).

By being exposed to contemporary social conditions that have links to oppressed populations in the past, students gain a deeper understanding of why such conditions exist today. In learning about how the historical past is connected to the present, students may become less judgmental of some of the reprehensible conditions existing in formerly colonized nations and take another look at their causes. The stereotypical questions and statements—Why don't they [the people of the country in question] pull themselves up by their own bootstraps? or They are lazy [often uttered about starving populations in African nations]—will be cast aside as students ask more complex questions as a result of their deeper, wider knowledge base.

Teaching Current Events from a Global Perspective

A fourth pedagogy in bringing a global awareness to requisite social studies courses in schools involves integrating current events in social studies classes. It is my strong opinion that social studies professionals have a moral responsibility to expose our students to contemporary affairs and their effects on the United States and world community (Kirkwood-Tucker, 1999). Discussion of what is happening in the world is one of the primary functions of social studies education. The systematic integration into instruction of what is happening in the contemporary world draws students into complex and challenging issues confronting communities and nations everywhere. Contemporary events can be examined in print and nonprint sources by utilizing local, state, national, and international media representing a variety of political views. At the conclusion of the semester, I require students to determine whether the news media prioritize or omit entirely specific world affairs in their reports.

Applying the Hanvey conceptual framework. Throughout my thirty-year career as a global educator I have used the conceptual framework developed by Robert Hanvey (1976) to provide organizational structure to the teaching of content from a global perspective, which students have to emulate when reporting current events to class (see Kirkwood-Tucker, 2001b). The framework's five dimensions include state of the planet awareness; perspective consciousness; cross-cultural awareness; knowledge of global dynamics; and awareness of human choices. I have chosen the Syrian uprising to demonstrate the application of this framework.

The first dimension, the state of the planet awareness, is defined as an "awareness of prevailing world conditions and developments including emergent conditions and trends, e.g., population growth, migration, economic conditions, resources and physical environment, political developments, science and technology, law, health, inter-nation and intra-nation conflicts, etc." (Hanvey, 1976, p. 5). Standing in front of my large freestanding world map, my faithful companion whenever I teach, I tell the class: "Today, I am reporting from Damascus, Syria, in the Middle East, where a large number of ordinary civilians are demanding political, cultural, and economic rights from their autocratic government. So far, over fifteen thousand civilians, soldiers, and dissidents have been killed and tens of thousands wounded in brutal crackdowns by President Bashar al-Assad. The military has strict orders to kill anyone demonstrating in the streets. Despite international outcry, Syria receives support from China and Russia and has refused to abide by the recommendation of the Arab League to a meeting between rebels and the government to stop the violence."

After the content of the current event is presented, questions to students are based on the facts of the event (including geographic concepts if time permits) to test student comprehension. Teacher: "Now, I will frame questions that apply to the state of the planet awareness dimension of our conceptual framework to test if you have paid attention."

- What is happening in Syria?
- Who can summarize the who, what, where, when, and why of this event?
- What is the name of Syria's president?
- Who can guesstimate the purpose of the Arab League?
- What is the role of Russia and China in Syria?

- Where is Syria located?
- What is the name of its capital?
- What is the capital's latitudinal and longitudinal position?
- On which continent and in which hemisphere is Syria positioned?
- How many time zones is Damascus ahead of the U.S. capital?

Teacher: "I now will address the second dimension of the conceptual framework, perspective consciousness, and how this dimension applies to the Syrian uprising." The perspective consciousness dimension is defined as "the recognition on the part of the individual that he or she has a view of the world that is not universally shared, that this view of the world has been and continues to be shaped by influences that often escape conscious detection, and that others have different views of the world" (Hanvey, 1976, p. 3). For this dimension, the teacher probes the belief systems of students as well as the agencies involved in the event, reminding students that individuals, groups, and nations hold different views of the world, which must be acknowledged and respected but not necessarily condoned. The teacher asks:

- What different beliefs are demonstrated in this uprising?
- What is your opinion of the cruel crackdown on the Syrian people by its government?
- How might the Syrian government perceive the alleged oppression of its people?
- What might other nations including the United States think of the uprising?
- Why do China and Russia oppose international interference in the uprising?

The teacher continues: "Now let's examine the cross-cultural awareness dimension of our conceptual framework and how it applies to this current event." Cross-cultural awareness is defined as an "awareness of the diversity of ideas and practices to be found in human societies around the world, of how such ideas and practices compare, including some limited recognition of how the ideas and ways of one's own society might be viewed from other vantage points" (Hanvey, 1976, p. 8). In this dimension, the teacher focuses on the commonalities of and differences between the United States and other nations and how our country might be viewed by other nations. Cultural practices range from dating practices, marriage customs, celebration of holidays, eating styles, language(s) spoken, form of government, religion(s), and education of boys and girls. In this dimension, the concept of empathy for the Other is emphasized. Examples of questions to pose include the following:

- What commonalties do we share with Syria? What differences exist?
- What are some of the cultural practices in Syria?
- What is your reaction to the knowledge that thousands have been killed and tens of thousands wounded by the military since the uprising?
- Are the Syrian people different from us? What are they seeking?
- Could Syria be described as having a culture of oppression?

Teacher: "Now let's move on to examine the fourth dimension of the conceptual framework: knowledge of global dynamics, and how it applies to the Syrian uprising." Knowledge of global dynamics is defined as "some modest comprehension of key traits and mechanisms of the world system, with emphasis on theories and concepts that may increase intelligent consciousness of global change" (Hanvey, 1976, p. 15). This dimension looks for examples from students about how the world is interconnected as a result of the event and probes for unanticipated consequences of the event. I engage students in examining the effects these interconnections have on their own country, schools, communities, and individual lives. The teacher challenges the students:

- What are the effects of this uprising on Syrian communities? On families? On the country? Neighboring countries? The world?
- Who is particularly affected by this crackdown in Syria? What might be unanticipated consequences of the crackdown on the regime? On rebels? On neighboring countries? On the United States?
What might be the effects of the uprising on teachers, students, and schools in Syria? Neighboring countries? The United States?

The teacher continues: "We now have reached the last dimension of our conceptual framework: awareness of human choices. Let's think about what questions we could ask to meet the definition of this dimension." Awareness of human choices is defined as "some awareness of the problems of choice confronting individuals, nations, and the human species as consciousness and knowledge of the global systems expand" (Hanvey, 1976, p. 26). The teacher probes students on the choices made in the event by individuals/groups/governments, knowing how choices can shape the lives of people, communities, nations, and the future of the world in significant ways. I continue challenging the students to make decisions to address this issue. I ask them what type of action they would take (if any), challenging them to develop activities or projects they could become engaged in to work towards a more peaceful, sustainable world. The teacher might ask the following:

- What choices were made in this uprising? Who made them?
- Who made the right choices? Is there a "right" choice?
- What choices would you make if you were the government? The civilians? What can we do to show our empathy to the Syrian civilians? The military? The dissidents?

In discussing the Syrian uprising, surely many more questions can be asked if teachers' time constraints were not a problem. It is very important for teachers to understand that not all questions have answers or that we do not always know the answer to questions we are posing. The main idea in applying this conceptual framework is to ask challenging questions to make students think critically and reflect about the larger world and its people and plant seeds in their minds as to the ramifications of current events on individuals, groups, and entire nations. It is hoped that exposure to the larger world will lead students to intellectual curiosity that makes them want to gain more knowledge and understanding about the world.

In schools, students could report on current events from a global perspective every Monday or Friday throughout the semester. Forming teams of five students each would ensure that each team member is able to cover one dimension of the conceptual framework from a global perspective. I also recommend teachers teach this conceptual framework at the beginning of the semester and have students apply one, two, three, or all dimensions of the framework to current events as well as their understanding of textual materials used in the classroom. If time allows, teachers may consider including basic geography skills to improve students' geographic knowledge, which critics of social studies education repeatedly refer to as inadequate (Levin, 2005).

At Florida State University, I have made the integration of current events from a global perspective a regular weekly assignment in all methods classes, requiring individual students, pairs, or teams to report on current events at the beginning of each class throughout the semester. By the end of the semester, teacher candidates have internalized the conceptual framework, empowering them to teach content from a global perspective.

Simulating the United Nations

My fifth pedagogy in bringing a global awareness to required social studies curricula and instruction embraces the simulation of the United Nations General

Assembly—an all-embracing interactive methodology that offers teachers and students the opportunity to examine the purpose and function of this international peace-building forum as well as to explore three of its seminal documents, the Universal Declaration of Human Rights (UDHR), the Convention of the Rights of the Child (CRC), and the Millennium Development Goals and Targets (United Nations, 2010) (see Kirkwood-Tucker, 2004).

In this simulation, student ambassadors from industrialized and nonindustrialized nations including the United States examine permeating human rights violations in their nation. Assigned readings can make a profound impact on students' acquiring a broad knowledge and understanding about the larger world and its conditions—for example, the discovery that in Canada 5.3 children per 1,000 live births die annually, whereas in Afghanistan the infant mortality rate is 154 per 1,000 live births (Population Reference Bureau, 2010). Rich dialogue emerges from ambassadors' multiple perspectives on new insights gained from grueling conditions, as many student ambassadors are unaware of the wide socio-economic differences in countries around the world.

Participating countries in the United Nations General Assembly should be selected equally from the industrialized and nonindustrialized world including the five permanent members of the Security Council: the People's Republic of China, the Republic of France, the United Kingdom, the Russian Federation, and the United States of America. The General Assembly sessions must be conducted with a formal protocol and should not be interrupted with trivia. Ambassadors must use formal (diplomatic) language in all conversations, debates, and resolutions before the General Assembly and with each other. Ambassadors are required to address their counterparts as "Distinguished Ambassador of X." The Chair of the General Assembly, role-played by the teacher to ensure a democratic learning community, must be addressed as the "Distinguished Chair of the General Assembly."

Sessions before the United Nations General Assembly include ambassadors' positions on assigned materials concerning conditions around the world including in their own country; the United Nations Millennium Goals (United Nations, 2010); analysis of the UDHR and the CRC and their violations interspersed with the latest news reports; and ambassadors' formal resolutions on the most pressing global issues needing immediate attention. Resolutions can range from topics on child labor; child soldiers; global security; HIV/AIDS; hunger and malnutrition; effects of nongovernmental organizations; poverty; status of women; and terrorism, among others. Focus of the last session of the General Assembly is a debriefing of the course and celebrating the closing ceremony with authentic culinary delights and national anthems.

Creative Pedagogies   251

In schools, the simulation can be implemented either throughout a semester or nine-week grading period, reserving Mondays or Fridays for the activity, or during a rigorous two-week period at the conclusion of a semester. I know of dedicated teachers who train students after school to participate in district or regional Model United Nations simulations.

At Florida State University, we have implemented a three-credit-hour course, Teaching Global Issues Simulating the United Nations, required at both the undergraduate and graduate levels, permitting us to simulate the United Nations General Assembly weekly for the entire semester. Assigned reading materials challenge ambassadors to have dynamic and controversial discussions.[3] Two sessions need to be reserved for ambassadors to present their unit plans, which must include a lesson on the United Nations, the Universal Declaration of Human Rights, and the Convention of the Rights of the Child as well as one or more global issues of their choice.

During the sessions, an efficient seating arrangement for the members and Chair of the General Assembly conducive to equal spatial representation and clear visibility of each other (a square) is critical. The formal structure of the General Assembly classroom provides an ideal forum to debate controversial issues and practice decision making in which participants can transcend their ordinary student roles to engage in democratic processes while searching for solutions to worldwide conditions. The configuration is also intended to create an atmosphere of intimacy. The splendor of the ambassadors' authentic attire, consisting of shawls I have collected from around the world, enhances the ambience of a converted classroom. I address my students as the "Distinguished Ambassador of X," a formality they immensely enjoy. The teacher acting as Chair of the General Assembly assumes a background role, mainly to maintain time constraints and ensure dynamic flow of the classroom discourse, as ambassadors are the central actors of the simulation.

Conclusion

Over seventy years ago Dewey (1939) defined citizenship competence as "the will to inquire, to examine, to discriminate, to draw conclusions on the basis of evidence after taking pains to gather all available evidence" (p. 31), claiming that "this intelligence gives students the power to see the world for what it is and reflect upon what it might become" (Dewey, 1963, p. 298). Dewey foreshadowed the purpose of global education in enhancing global citizenship competence in his vision of youngsters becoming well-informed, rational, humanistic, and critical thinkers in a challenging, competitive world.

Global educators in teacher education programs and schools in the United States and worldwide are continuously searching for creative pedagogies to empower teacher candidates and classroom teachers in globalizing curriculum and instruction. Teachers will joyfully discover that exploring the larger world in their classrooms leads to student excitement and student intellectual curiosity. Since the teacher is the ultimate curricular and instructional gatekeeper, it is my strong conviction that if a better world can be created, it must begin with teachers.

Notes

1. The terms "teacher candidates," "prospective teachers," and "future teachers" are used interchangeably.
2. At this point teachers may wish to distinguish between the concepts of uprising, intra-nation and inter-nation conflict, civil war, and revolution.
3. *A Global Agenda: Issues before the United Nations.* Published annually by the United Nations Association of the United States, Korea University and the World Federation of United Nations Association, New York. This book covers the annual fall agenda of the United Nations General Assembly beginning in September and is available in August.

    60 Ways the United Nations Makes a Difference. (2006). DVD published by the United Nations Department of Public Information.

    The United Nations Today. (2008). United Nations Publication Sales No. E.08.I.6, United Nations Headquarters, http://www.un.org. This book provides in-depth information on the historical development of the United Nations, its organizations, and mission.

    World Data Sheet. Published annually by the Population Reference Bureau, Washington DC. This large chart provides the latest facts about conditions in every nation of the world. http://www.populationeducation.org

References

Ames, A. (2011, June 5). Infant mortality: Racial disparities defy easy explanations. *Tallahassee Democrat.*
Cushner, K., & Brennan, S. (2007). *Intercultural student teaching: A bridge to global competence.* Lanham, MD: Rowman & Littlefield.
Dewey, J. (1939). *Theory of valuation.* Chicago: University of Chicago Press.
Dewey, J. (1963). *Philosophy and civilization.* New York: Putnam.
Grahame, K. (1908). *The wind in the willows.* New York: Scribner.
Hanvey, R. G. (1976). *An attainable global perspective.* Denver, CO: Center for Teaching International Relations.
Heater, D. (2002). Does cosmopolitan thinking have a future? *Review of International Studies, 26,* 179–197.
Kirkwood-Tucker, T. F. (1999, November/December). Reporting the world: Teaching current events from a global perspective. *Social Studies and the Young Learner, 12*(2), 29–31.
Kirkwood-Tucker, T. F. (2001a). Building bridges: Miami "ambassadors" visit Russia. *Social Education, 65*(4), 236–239.

Kirkwood-Tucker, T. F. (2001b, Winter). Preparing teachers to teach from a global perspective. *The Delta Kappa Gamma Bulletin*, 67(2), 5-12.

Kirkwood-Tucker, T. F. (2002, Winter). Teaching about Japan: Global perspectives in teacher decision-making, context, and practice. *Theory and Research in Social Education*, 30(1), 88-115.

Kirkwood-Tucker, T. F. (2003). Global perspectives teaching in the elementary classroom: Reflections on my practice. *The International Social Studies Forum*, 3(1), 285-290.

Kirkwood-Tucker, T. F. (2004, Winter). Empowering teachers to create a more peaceful world through global education: Simulating the United Nations. *Theory and Research in Social Education*, 32(1), 56-74.

Kirkwood-Tucker, T. F. (2009). From the trenches: The integration of a global perspective in curriculum and instruction in the Miami-Dade County public schools. In T. F. Kirkwood-Tucker (Ed.), *Vision in global education: The globalization of curriculum and pedagogy in teacher education and schools* (pp. 137-162). New York: Lang.

Kirkwood-Tucker, T. F., & Benton, J. (2002). The lessons of Vietnam. *Social Education*, 66(6), 362-367.

Levin, M. H. (2005). Putting the world into our classrooms: A new vision for 21st century education. PPI online. Retrieved from http:// www.ppionline.org/ppi_ci.cfm?knlgAreaID=110&subsecid

Merryfield, M. M. (1997). A framework for teacher education in global perspectives. In M. M. Merryfield, E. Jarchow, & S. Pickert (Eds.), *Preparing teachers to teach global perspectives: A handbook for teacher education* (pp. 1-24). Thousand Oaks, CA: Corwin Press.

Merryfield, M. M. (2009). Moving the center of global education: From imperial world views that divide the world to double-consciousness, contrapuntal pedagogy, hybridity, and cross-cultural competence. In T. F. Kirkwood-Tucker (Ed.), *Visions in global education: The globalization of curriculum and pedagogy in teacher education and schools* (pp. 215-239). New York: Lang.

Myers, J. P. (2006). Rethinking the social studies curriculum in the context of globalization: Education for global citizenship in the U.S. *Theory and Research in Social Education*, 34(3), 370-394.

Population Reference Bureau. (2010). *2012 world data sheet*. Retrieved from http://www.prb.org

Quezada, R., & Cordeiro, C. (2007). Biliteracy teachers' self-reflections of their accounts while student teaching abroad: Speaking from "the other side." *Teacher Education Quarterly*, 43(1), 95-113.

Thornton, S. (2005). *Teaching social studies that matters: Curriculum for active learning*. New York: Teachers College Press.

United Nations. (1948). *Universal declaration of human rights*. Retrieved from http:// www.un.org/en/documents/udhr/

United Nations. (1990). *Convention on the rights of the child*. Ratified 20 November 1989 and put into force 2 September 1990. Retrieved from http:// www2.ohchr.org/ english/law/crc.htm

United Nations. (2010). *Millennium development goals and targets*. Retrieved from http:// un/org/millennium/declaration/ares552e.htm

CHAPTER TWENTY-ONE

Social Studies Education in a Globalized Era: Afterword

Lydiah Nganga
John Kambutu

Globalization is a not a new phenomenon. Rather, the world has always experienced events that facilitated economic, political, cultural, social and military interconnectedness. The current magnitude and the effect of contemporary globalization, however, are new phenomena. Indeed, the uniqueness of contemporary globalization is perhaps evident in the transformation of the world into a "village." The ongoing advances in technology and information are perhaps responsible for the emerging global village. Due to advances in technology, for example, the physical and cultural barriers that previously hindered global interactions are no longer a reality. Instead, improved technologies are allowing people of different cultural persuasions to interact both physically and virtually at rates never imagined, thus turning the world into a place of interdependence and interconnection. Due to global interconnection and interdependence, the effects of an event in one part of the world are now reverberating throughout the entire world (Wiarda, 2007). The ongoing global economic crisis is a good example of how well the world is interconnected.

Given that the world is now a web of complex interconnections, educators ought to rethink social studies education. To prepare for the complexities inherent in a globalized period, the National Council for the Social Studies (NCSS, 2010) recommends an education that equips learners with skills, knowledge and dispositions essential for global citizenship. While skills in technology and informational management are essential, preparing learners for effective functioning in a world that is interconnected economically, socially, culturally and, to a large extent, politically is equally important. Meanwhile, as the world shrinks into a village, there is an urgent need to develop skills in diversity appreciation and environmental justice. As a result, the NCSS has continued to support a social studies education with a global competency focus.

Implementing an effective social studies education is potentially a daunting task. For example, rapid changes in technology and information necessitate an ongoing review of content in order to provide a valid curriculum. Meanwhile, increased global interactions, due to increased interconnectedness and interdependence, creates a need for educators to be well informed about global issues and events. To that end, social studies education is likely to be effective when it is designed around themes and units with diverse global perspectives (NCSS, 2001a). Thus, in its position statement on global education, the NCSS encourages educators to help students to understand the existing interconnectedness of human and natural environments, including the interrelated nature of global events and issues. However, because globalization is understood and interpreted differently, it is almost impossible to develop a comprehensive social studies education (Haugen & Mach, 2010). Notwithstanding the available divergent views relative to globalization, educators should always ensure that the social studies curriculum is accurate and objective.

An accurate social studies education entails a holistic exploration of globalization. For example, a broad and inclusive education should always address both the positive and negative aspects of globalization. Positively, globalization might be discussed in the context of the benefits of expanding global capitalism. While beneficiaries of expanding global capitalism are likely to favor globalization, the groups that are dominated and negatively affected by globalization are likely to view it negatively. To the dominated groups, globalization is simply a scheme to expand the economic, political and cultural hegemonies of wealthy nations of the West (Agbaria, 2011a). Meanwhile, ideological competition between stakeholders such as policy makers, politicians and activists could complicate further the process of teaching objectively (Evans, 2004; Ross, 2001; Westheimer & Kahne, 2004). Notwithstanding the challenges involved, however, there exist multiple global education models that social studies educators could utilize in their efforts to revise curricula. To be sure, existing models have different emphases and interpretations, but a model that advocates an education for global outlook, consciousness and citizenship is highly recommended (Hanvey, 1972; Hicks, 2003; Kniep, 1987). However, it is also crucial for educators to develop consensus relative to instructional terminologies.

There are several terminologies that tend to be used in a social studies global curriculum. For example, while global education, world-centered education and global perspectives are applied explicitly, terminologies that address issues of human rights, economic and environmental justice, international conflicts and the effects of global technology and informational changes are typically implied. However, because social studies teachers have a responsibility to prepare learners for global citizenship, they must ensure that all essential skills and values are explicitly included in social studies (NCSS, 2010). Realistically, such a task is not easy. Consider, for example, the ongoing discussions in the United States relative to curriculum challenges, scale and approaches to ensure that students are prepared adequately for post-9/11 global challenges (Agbaria, 2011b).

Challenges and Solutions

In this book, educators have identified several challenges that they have experienced while teaching for global education. In addition, they have provided possible practical solutions in the form of lesson plans, instructional activities and perspectives. In particular, there is a call for educators to be imaginative and creative while exploring and discovering global realities. The use of activities and materials that help learners to connect local, national and global occurrences is especially critical because they enable learners to familiarize themselves with different aspects of global interconnectedness. In addition, the NCSS (2001b) favors the adoption of an interdisciplinary curriculum approach both within and beyond social studies. A curriculum that links directly to multicultural education is needed because it promotes the study of global issues from the dimensions of intellectual honesty, multiple perspectives and social justice. Additional effective instructional approaches include the use of diverse primary resources such as models, artifacts, literature and guest speakers from other countries. Equally beneficial is the use of advanced technologies.

When selected and applied properly, technology is capable of immersing learners in thought-provoking "virtual visits." Some technologies, such as Google Maps, allow learners to virtually "visit" and experience other parts of the world without ever leaving their physical classrooms. Meanwhile, interactive instructional methodologies such as cross-cultural simulations and role-playing are other relevant approaches to teaching global education. But doing objective research is another ideal way to develop informative global pedagogy. Thus, the NCSS encourages teachers not only to become researchers, but also to make sure that curriculum and instructional processes are research based (NCSS, 2010b).

Teaching social studies education in an age of globalization has many challenges. However, because diverse teaching resources are available, committed educators are most likely to find relevant instructional resources. In a globalized age, it is essential for social studies education to focus on preparing learners for global citizenship. A variety of skills, knowledge and dispositions are needed, but having a global consciousness and the ability to consider the world from multiple perspectives is a necessity. Equally important to develop are social justice skills. In all, then, teaching for global citizenship requires creativity, resourcefulness, endurance and a critical awareness of global dynamics. Although skills in cross-cultural understanding and problem solving are necessary, educators with a broader outlook of world issues, including the different ways that globalization affects different cultures and groups on the planet, might implement global education successfully.

Finally, social studies educators have a responsibility to develop curricula that elucidate different levels of global perspectives. Therefore, they must be fully prepared to implement global education in the context of globalization.

References

Agbaria, A. K. (2011a). Debating globalization in social studies education: Approaching globalization historically and discursively. *Intercultural Education*, 22(1), 69-82.

Agbaria, A. K. (2011b). The social studies education discourse community on globalization: Exploring the agenda of preparing citizens for the global age. *Journal of Studies in Intercultural Education*, 15(1), 54-74.

Evans, R. W. (2004). *The social studies wars: What should we teach the children?* New York: Teachers College Press.

Hanvey, R. G. (1972). *An attainable global perspective*. New York: Center for Global Perspectives in Education.

Haugen D., & Mach, R. (2010). *Opposing viewpoints: Globalization*. New York: Gale, Cengage.

Hicks, D. (2003). Thirty years of global education: A reminder of key principles and precedents. *Educational Review*, 55(3), 265-275.

Kniep, W. M. (1987). *Next steps in global education: A handbook for curriculum development*. New York: American Forum.

National Council for the Social Studies. (2001a). *Preparing citizens for a global community*. Retrieved from http://www.socialstudies.org/positions/global

National Council for the Social Studies. (2001b). *What are global and international education?* Retrieved from http://www.socialstudies.org/positions/global

National Council for the Social Studies. (2010). *National curriculum standards for social studies: A framework for teaching, learning, and assessment*. Silver Spring, MD: NCSS Task Force.

Ross, E. W. (2001). *Power politics*. Cambridge, MA: South End Press.

Westheimer, J., & Kahne, J. (2004). What kind of citizen? The politics of educating for democracy. *American Educational Research Journal*, 41(2), 237-269.

Wiarda, H. J. (2007). Globalization in its one and many forms. In H. J. Wiarda (Ed.), *Globalization: Universal trends, regional implications* (pp. 264-276). Lebanon, NH: University Press of New England/Northeastern University Press.

CONTRIBUTORS

Rebecca C. Aguayo is a graduate student in Social Studies Education at the University of Missouri (MU). Prior to graduate school, she taught secondary history for four years. Her research interests include human rights abuses and genocide education, curriculum policy, and global citizenship.

Linda Bennett is a Special Assistant to the Provost at the University of Missouri (MU). In 2011-2012, she served as an American Council on Education (ACE) Fellow. Bennett received her doctorate in Education, with an emphasis in elementary education, from the University of Northern Colorado. She has given more than ninety presentations on educator training and higher education in the United States, Canada, Mexico, Australia, and Switzerland. Dr. Bennett has published more than thirty international and national articles, seven book chapters, and *Digital Age: Technology-Based K-12 Lesson Plans for Social Studies*. She was selected as the editor for *Social Studies and the Young Learner* (2006-2010) and was the inaugural inductee into the University of Memphis Education Hall of Fame. MU honors include the Excellence in Education Award, Excellence in Teaching with Technology Award, and Graduate Professional Council Gold Chalk Award.

Dorothy Blanks was raised in Washington State and Seoul, Korea, and earned a BA in history from the University of Washington. Her master's degree is in Secondary Education from the University of Guam, and she is now completing a PhD in Teacher Training in the Social Studies from the University of Tennessee. Dorothy taught middle school social studies and language arts for twelve years in Budapest, Hungary; Moscow, Russia; Taipei, Taiwan; and Guam. Her primary interests include interdisciplinary studies, global education and citizenship, and constructivist methods of teaching social studies.

Aaron T. Bodle is an assistant professor at James Madison University, where he teaches social studies methods courses for elementary educators. He recently received his doctorate in Curriculum, Instruction, and Teacher Education from Michigan State University. His research explores changing conceptions of citizenship education in a globalized world; students' perceptions of their connections to "global," "national," and "local" contexts; and links between multicultural and global education in theory and practice in K-12 social studies.

Kenneth T. Carano is an assistant professor in the Division of Teacher Education with an emphasis in Social Studies Education at Western Oregon University. He is also a returned Peace Corps volunteer who spent two years teaching elementary students and running an after-school program in Suriname in South America. His research interests focus on global perspectives in teacher education programs and preparing students to be effective citizens in a world that is becoming increasingly interconnected. He has authored numerous articles on global education and was a short story laureate for one of his works of fiction.

Antonio J. Castro is an assistant professor at the University of Missouri (MU), Department of Learning, Teaching, and Curriculum, 211F Townsend, Columbia, MO 65211; castroaj@missouri.edu. His research interests include the recruitment, preparation, and retention of teachers for culturally diverse contexts and urban schools as well as multicultural and global citizenship and democratic education. He has authored and co-authored numerous articles in journals such as *Action in Teacher Education*, *Social Studies and the Young Learner*, and *Theory and Research in Social Education*, among others.

Catherine Cooke-Canitz is a doctoral student in Social Studies Education and Policy at the University of Missouri-Columbia. Simultaneously, she has earned a certificate in International Development and is working on a TESOL endorsement for her elementary teaching license. Ms. Cooke-Canitz has roughly twenty years' experience in urban and rural public elementary schools, during which she received her National Board Teacher Certification. She currently teaches elementary social studies methods to pre-service teachers. Her research interests include the acculturation of teachers to foreign students and issues of social justice, including how Native Americans are addressed in history standards across the country, and the critical teaching of history and patriotism in the early elementary years.

Jason L. Endacott is Assistant Professor of Secondary Social Studies Education at the University of Arkansas-Fayetteville. Dr. Endacott received his PhD (2007) and MS (2001) in Curriculum and Instruction from the University of Kansas and his BS (1998) in Education from Kansas State University. He has published articles in *Theory and Research in Social Education*, *The Social Studies*, and *Social Studies Research and Practice*, and has disseminated the results of his work at annual conferences for the National Council for the Social Studies, College and University Faculty Assembly, and International Society for the Social Studies.

Contributors

Paul G. Fitchett is Assistant Professor of Education in the Department of Middle, Secondary, and K12 education at the University of North Carolina–Charlotte. His scholarly interests include social studies education, culturally responsive teaching, teacher workplace conditions, and educational policy. Dr. Fitchett's research has been published in numerous journals including *Theory and Research in Social Education, Action in Teacher Education, Urban Education,* and *Journal of Social Studies Research.* He can be contacted at Paul.Fitchett@uncc.edu

Erik Garrett teaches in the Department of Communication & Rhetorical Studies at Duquesne University. His doctorate is from Purdue University, where he achieved a joint degree in both Philosophy and Communication. Dr. Garrett received the 2010 Duquesne University Teaching Award. He recently published in 2011. "The Rhetoric of Anti-Black Racism: Lewis R. Gordon's Radical Phenomenology of Embodiment" in the *Atlantic Journal of Communication* (19(1), 6-16). His research interests are phenomenology, philosophy of communication, urban communication, rhetoric, social political philosophy, environmental justice, and globalization. Dr. Garrett has just completed a manuscript on zoo animals, rhetoric, and phenomenology.

Rosanna M. Gatens received a PhD in Modern European History from the University of Pittsburgh. She is a specialist in the social and intellectual history of Germany, especially during the era of the Weimar Republic and the Third Reich. She writes about the collapse of academic freedom in German universities during the 1920s, the anti-fascist and anti-militarist campaigns of the German League for Human Rights during the interwar years, and racism as a component of National Socialism. Her research and writing in the area of Holocaust education focus on the effectiveness of particular teaching and learning strategies for high school and college students. She has planned and conducted seminars, symposia, and workshops on Holocaust education since 1982. She is currently Director of the Center for Holocaust and Human Rights Education in the College of Education at Florida Atlantic University. She received the 2008 Human Relations Award from the American Jewish Committee of South Florida and the Florida Atlantic University TIAA-CREFF Community Service Award in 2011. She is a member of the Florida Task Force on Holocaust Education and a member of the Save Darfur Coalition of South Palm Beach.

Keonghee Tao Han is an Assistant Professor at the University of Wyoming–Casper College Center. She earned her PhD in Literacy Studies and a TESOL Masters in Literacy Studies from the University of Nevada-Reno. Her research interests include qualitative descriptions of English Learners' academic and social aspects of literacy learning in K–8 classrooms, pre-service teachers' dispositions towards social justice education, and diverse faculty and students' lived experiences in remote educational settings. Han has published in journals such as *International Journal of Progressive Education*, *Journal of Educational Practices for Social Change*, and *Urban Education*.

Jason R. Harshman is a doctoral candidate and Lecturer in Social Studies and Global Education at The Ohio State University. He has received travel grants to study in South Korea, Japan, and Turkey. His research interests include examining the intersections of youth cultures and critical geography within global education, critical theory and pedagogy, and developing how media and technology are used to support global citizenship education. He has published in *Educational Forum* and the *Journal of the Research Center for Educational Technology*, and is co-editing a book on research in global citizenship.

Emma Kiziah Humphries is the Assistant in Citizenship at the Bob Graham Center for Public Service at the University of Florida (UF) in Gainesville, working to implement a three-year, $3 million Knight Foundation Grant to prepare UF students to be informed, skilled, and engaged citizens. She also teaches courses in social foundations in UF's College of Education, from which she earned her PhD in Curriculum and Instruction. Her primary research interests include civic education in secondary and undergraduate education and the intersection of civic engagement and social technologies.

Mirynne Igualada received her MEd in Culture, Curriculum, and Educational Inquiry at Florida Atlantic University and is now a doctoral student in the department of Culture, Curriculum, and Educational Inquiry. She served as a social studies teacher for eight years before becoming the Advanced Studies Coordinator for Broward County Public Schools. She focuses on coordinating, developing, and monitoring the curriculum for the district's Advanced Placement program. Her scholarly interests include how global education can provide an opportunity for both teacher and student empowerment through curriculum and international experiences. She has written, reviewed, and implemented lesson plans for Pearson and McGraw-Hill.

John Kambutu is Associate Professor of Educational Studies at the University of Wyoming–Casper College Center. He received his PhD in Education from the University of Wyoming. His research work is in cultural diversity, rural education, and transformative learning. He also has scholarly interest in globalization/internationalization efforts. Dr. Kambutu has published several articles and book chapters. In addition, he has guest-edited a special themed issue entitled "Multicultural Education in an Internationalized/Globalized Age" in *Multicultural Perspectives: The Official Journal of the National Association for Multicultural Education*. An award-winning educator, John Kambutu believes strongly in an education that liberates humanity from the ills of ignorance, thus enabling all to live a free life. An education for critical change is transformative in design and practice. You can contact him at Kambutu@uwyo.edu.

Toni Fuss Kirkwood-Tucker is Associate Professor Emerita of Florida Atlantic University and now serves as Visiting Associate Professor in Social Science and Global Education and Graduate Coordinator in the School of Teacher Education at Florida State University. She is German American of bicultural multilingual background, was a Fulbright Scholar to China and Russia, and was the recipient of the Distinguished Global Scholar Award given by the International Assembly of the National Council of the Social Studies. Toni's primary research interests lie in the criticality of balance of global issues in curriculum and instruction and multiplicity of approaches in the integration of a global perspective in teacher education programs and schools. Contact her at tkirkwoodtucker@comcast.net.

Lance McConkey is a PhD candidate in the department of Teacher Education at the University of Tennessee. He received his BS in History from Tennessee Wesleyan College and his master's degree in Education from Lee University, as well as an EdS from the University of Tennessee in Teacher Education. Lance currently teaches high school social studies in a rural school system in Tennessee. He has written several articles addressing social studies education and has served as a co-author for various book chapters.

Lydiah Nganga is Associate Professor of Elementary and Early Childhood Education at the University of Wyoming–Casper College Center. She teaches Humanities/Social Studies Methods and Early Childhood courses. Dr. Nganga earned her PhD in Curriculum and Instruction from the University of Wyoming. Her research focuses on global/international education, curriculum studies, multicultural education and social justice. Dr. Nganga has published two books (one a children's book). She has also authored/co-authored numerous articles and chapters that have been featured in books such as *Critical Pedagogy and Teacher Education in the Neoliberal Era* (S. L. Greonke & J. A. Hatch, eds.); *Curriculum Development: Perspectives from Around the World* (J. D. Kirylo & A.K. Nauman, eds.); *Early Childhood Education in Rural Commu-*

nities: *Access and Quality Issues* (D.T. Williams & T.L. Mann, eds.); and journals such as *Teaching and Teacher Education; Spaces for Difference: An Interdisciplinary Journal*; and *Early Years: An International Journal of Research and Development* among others. Nganga is also an award-winning educator. Contact her at Lnganga@uwyo.edu.

Joe O'Brien is Associate Professor in Middle/Secondary Social Studies at the University of Kansas. He was on the faculty at Virginia Commonwealth University, earned a BA degree in history and a MSE and EdD degree at the University of Virginia, and taught social studies in grades 7-12. His research interests include history teachers' use of primary sources and the instructional use of social media. He has published work in journals such as *Contemporary Issues in Technology & Teacher Education* and *Social Education* and has received numerous awards for teaching.

Sarah Lewis Philpott is a PhD candidate in the department of Teacher Education at the University of Tennessee. She received her BS in Interdisciplinary Studies from Tennessee Wesleyan College and both her MEd and EdS from Lincoln Memorial University. Sarah is a former elementary and middle grades teacher, has written professional articles for journals such as *Social Studies and the Young Learner*, and has served as co-author for various book chapters. She is currently involved in multiple research and writing projects at the University of Tennessee.

William B. Russell III is Associate Professor of Social Science Education at the University of Central Florida. He teaches social studies education courses and serves as the Social Science Education PhD coordinator. Dr. Russell serves as the director for the International Society for the Social Studies and is the editor of the preeminent journal in the field of social studies education, *The Journal of Social Studies Research*. His research interests include alternative methods for teaching social studies, pre-service teacher education, and teaching with film. Dr. Russell has authored/edited numerous books. Dr. Russell has also authored numerous peer-reviewed journal articles related to social studies education, which have been featured in journals such as *Action in Teacher Education, Social Education, The History Teacher, Journal of Social Studies Research, The Clearing House, Social Studies and the Young Learner,* and *The Social Studies.*

Dilys Schoorman received her PhD in Curriculum and Instruction at Purdue University. She is currently a Professor at Florida Atlantic University, and her scholarly interests emerge from the intersections of multicultural and global education, leadership, and teacher education, all explored within a framework informed by critical theory and social justice pedagogy. She has published articles in *Equity and Excellence in Education*, *Journal of Citizenship and Social Justice*, *Journal of Teacher Education*, *Teaching and Teacher Education*, *Journal of Multicultural Education*, and *Diaspora, Indigenous and Migrant Education*, among others. Her co-authored article in *Intercultural Education* was the most downloaded article for that journal in 2010.

Rachayita Shah recently received a Ph.D. in Curriculum and Instruction from Florida Atlantic University. Her doctoral research focused on teacher professional development for Holocaust and human rights education. Her research interests include multicultural and human rights education, teacher professional development, and TESOL and bilingual education. She currently teaches Introduction to Diversity for Educators at Florida Atlantic University, a course that focuses on integrating multicultural approaches to curriculum and instruction to serve the needs of diverse learners in U.S. schools.

Daniel W. Stuckart is Associate Professor at the City University of New York, Lehman College in the Bronx, New York City. He serves as the undergraduate and graduate academic adviser for social studies teacher candidates, preparing them to teach in diverse schools with massive influxes of international students from around the world. His research interests include the intersection of technology, urban education, and democratic practices. He has authored numerous articles and is co-author of the book *Revisiting Dewey: Best Practices for Educating the Whole Child Today* (2010).

Karen Thomas-Brown is an associate professor in the School of Education at the University of Michigan–Dearborn. She holds a PhD in Geography from the University of the West Indies (Mona) in Geography. Dr. Thomas-Brown has published over twelve articles in peer-reviewed journals in the United States and the Caribbean, and her research focus has been geography and multicultural education in a globalized setting. Her most recent publications include "Coping with Neoliberalism in Jamaican Towns: How Poor Jamaicans Use Informal Employments as a Means of Survival." Brown is a Commonwealth Scholar and a member of the Editorial Board for *Research in Geo-*

graphic Education. In addition, she has peer-reviewed articles for *Social Studies Research and Practice*.

Thomas N. Turner has been a Professor of Education at the University of Tennessee for over forty years. He has authored eight books and numerous professional articles. The fourth edition of his *Essentials of Elementary Social Studies* will be released this fall by Routledge. He teaches courses in social studies education, storytelling and drama, creative teaching, and trends and issues in education. His only grandchild, Tessa Nadia Turner, just turned five.

Ozum Ucok-Sayrak currently teaches in the Department of Communication and Rhetorical Studies at Duquesne University. She received her doctorate in Communication Studies from the University of Texas–Austin. Dr. Ucok-Sayrak has taught undergraduate and graduate courses in intercultural communication, interpersonal communication, small group communication, oral communication, public speaking, and business and professional communication, among other communication courses. Her research interests include the study of language and social interaction, ethnography of communication, health communication, and philosophy of communication. She is particularly interested in studying the construction of selfhood and identity through verbal and visual discourses. Dr. Ucok-Sayrak has published papers on how people interactively make sense of art, the aesthetic consequences of cancer treatment, and the transformations of self in surviving breast cancer in relation to changes in bodily appearance.

Julie Wachtel received her master's in Education from Long Island University specializing in Elementary Education and English Literature. She has been an elementary school teacher since 2000; her areas of expertise are writing and social studies education. In 2007, she began teaching for the University of Phoenix. There, she teaches college students the importance of reading and writing critically. During the summer of 2008 she co-authored an elementary school curriculum pertaining to the Holocaust and human rights, which has been implemented at her school since the fall of 2008. In 2009, she was honored with the Florida Atlantic University award for the Exemplary Holocaust Educator of the Year in Elementary Education. In 2011, she was awarded the Agnes Crabtree International Relations Award from the Florida Council for the Social Studies.

Elizabeth Yeager Washington is Professor of Social Studies Education at the University of Florida–Gainesville (UF), a Senior Fellow of the Florida Joint

Center for Citizenship, and a Knight Fellow at the Bob Graham Center for Public Service. She earned her PhD in Curriculum and Instruction from the University of Texas-Austin, previously served as editor of *Theory and Research in Education* (2001-2007), and currently coordinates the master's/certification program in Social Studies Education at UF, where she teaches secondary social studies methods, civics and government methods, and global studies methods courses. Her research interests include civic education in the middle school, teacher professional development in civic education, and the teaching and learning of history.

Stewart Waters is Assistant Professor of Social Science Education in the Department of Theory and Practice in Teacher Education at the University of Tennessee-Knoxville. His research interests include alternative methods for teaching social studies, character education, visual literacy, social studies curriculum, and teaching with film. Dr. Waters is the conference coordinator for the International Society for the Social Studies and is the assistant editor for *The Journal of Social Studies Research*. Dr. Waters has authored one book and several peer-reviewed journal articles related to social studies education that have been featured in journals such as *Action in Teacher Education*, *The Journal of Social Studies Research*, and *Social Studies Research and Practice*.

Cameron White is Professor of Social Education at the University of Houston. He received his PhD from the University of Texas specializing in Curriculum and Instruction/Social Studies Education. Research interests include social justice education, cultural studies and media, global/international education, community education, and social studies education. He has seven books and numerous articles in press and has served as co-PI (co-primary investigator) on seven Teaching American History grants. He can be reached at cswhite@uh.edu.

A.C. (Tina) Besley, Michael A. Peters,
Cameron McCarthy, Fazal Rizvi
General Editors

Global Studies in Education is a book series that addresses the implications of the powerful dynamics associated with globalization for re-conceptualizing educational theory, policy and practice. The general orientation of the series is interdisciplinary. It welcomes conceptual, empirical and critical studies that explore the dynamics of the rapidly changing global processes, connectivities and imagination, and how these are reshaping issues of knowledge creation and management and economic and political institutions, leading to new social identities and cultural formations associated with education.

We are particularly interested in manuscripts that offer: a) new theoretical, and methodological, approaches to the study of globalization and its impact on education; b) ethnographic case studies or textual/discourse based analyses that examine the cultural identity experiences of youth and educators inside and outside of educational institutions; c) studies of education policy processes that address the impact and operation of global agencies and networks; d) analyses of the nature and scope of transnational flows of capital, people and ideas and how these are affecting educational processes; e) studies of shifts in knowledge and media formations, and how these point to new conceptions of educational processes; f) exploration of global economic, social and educational inequalities and social movements promoting ethical renewal.

For additional information about this series or for the submission of manuscripts, please contact one of the series editors:

A.C. (Tina) Besley: t.besley@waikato.ac.nz
Cameron McCarthy: cmccart1@illinois.edu
Michael A. Peters: mpeters@waikato.ac.nz
Fazal Rizvi: frizvi@unimelb.edu.au

To order other books in this series, please contact our Customer Service Department:
  (800) 770-LANG (within the U.S.)
  (212) 647-7706 (outside the U.S.)
  (212) 647-7707 FAX

Or browse online by series:
  www.peterlang.com